Electricity, Magnetism, Gravity, & The Big Bang

♦ ♦ ♦

Charles R. Storey

ELECTRICITY, MAGNETISM, GRAVITY, & THE BIG BANG

SECOND EDITION

Copyright © Charles R. Storey 1999

International Standard Book Number 0-9638766-4-3
Printed in the United States of America
Copyright © 1999 by Charles R. Storey
All Rights Reserved
1999

ELECTRICITY, MAGNETISM, GRAVITY, & THE BIG BANG

I am dedicating this book to my lovely and beloved daughter, Tammy Jean Storey.

Forward

Many publications on this subject matter rely upon mathematics, descriptions and terminology which may be very technical and difficult to follow. In this book, you will find much of the descriptions greatly simplified and easier to understand. After you have read several chapters, you may come to the conclusion that the same underlying physical process is being described over and over again. This book is written in this manner to demonstrate that all of our laws of physics, including the principles of electricity, magnetism, and gravity, are a result of the same relatively simple, underlying physical process which is producing our perceptions about our universe in general and the world in which we live.

Contents

ACKNOWLEDGEMENTS

The information which is presented here was made possible by the hard work, tireless efforts, persistence, and determination of a great many researchers, physicists, mathematicians, computer programmers, and technical assistants. A special appreciation and thanks should be included for the contributions provided by Dale Hensel, and Robert McCartney for their informational and technical assistance which were instrumental in the development of this scientific overview.

Introduction

Have you ever stopped to wonder "Just what is the underlying physical process which is producing the effects of electricity, magnetism, and gravity?" What strange physical processes are at work within our universe to produce our perceptions of time and space? Why does acceleration and relative motion slow down clocks and shorten measuring rods according to Albert Einstein's Theory of Relativity? What does an atom really consist of? Does our universe go on forever, or does it have a finite size, and if so, what lies beyond its outer reaches? How can we use a Unified Field Theory to improve our means of space travel?

Throughout this book we will attempt to analyze these questions in such a way that the answers will soon become obvious to the reader. We will take a look at some of the early discoveries which led to the development of the theories regarding electricity, magnetism and gravity. Then, we shall see how more advanced theories were developed to include the latest observations, discoveries, and accumulating experimental evidence which have enabled us to progress to our current level of scientific achievement.

New theories lay the groundwork for new experiments, and new experimental evidence raises further questions about the nature of the universe in which we live. This is the learning process which is leading our civilization on a fantastic voyage of discovery to explore some of the most fascinating mysteries hidden within our universe.

NOTES

Chapter 1

THE ETHER THEORY

By the mid-1800's, the mathematics of Newtonian physics had been firmly established within the academic community in Europe. Recent advances in science and technology in the years immediately preceding this time period, gave researchers such a considerable amount of confidence that they now felt that their recently discovered scientific and analytical methods could be used to unlock the secrets of electricity and magnetism.

This task was undertaken primarily by a Scottish physicist named James Clerk Maxwell (1831 – 1879) with the assistance of Michael Faraday (1791 – 1867), Hans Christian Oersted (1777 – 1851), and several famous physicists of this particular time period. These physicists coordinated their research and shared large amounts of experimental information in an effort to formulate a comprehensive theory of electricity and magnetism.

Much of this early research on electricity and magnetism indicated that electric fields are some type of a "pressure" distortion, and that magnetic fields are a type of "change" distortion in some mysterious medium. Some of the details of this mysterious medium even began to show up in the mathematics which these early physicists were using to describe their theories.

As James Clerk Maxwell and his co-researchers continued to develop their theories concerning electricity and magnetism, they were able to predict the existence of electromagnetic waves and that light was produced by electromagnetic waves at certain frequencies. Due to the fact that waves of any type require a medium of some sort to conduct them, they further speculated that there had to be a mysterious gas or some other medium through which these waves travel or propagate.

During this time period, the exact nature of a gas was still a mystery to most physicists. Molecular theory had not yet been developed, and consequently, the manner in which a gas permeated space was still not fully understood. The presence of a gas could only be detected by the effects it produced, and by studying these effects, chemists were able to isolate and study several new gasses which had not been discovered previously.

Among the most interesting and mysterious of these was a gas called "ether". Ether was used as early as 1842 to anesthetize patients during operations, and since this gas was able to produce the almost unbelievable effects of making people unconscious, this was the mysterious gas that everyone was talking about at that time. Thus, many physicists began to suspect that there may be some other gas which permeated all of space, which had not been discovered previously, that was producing the mysterious effects of electricity and magnetism.

The physicists who were studying the effects of electricity and magnetism soon began to refer to this suspected gas as an "ether", or "mysterious gas". In 1873, James Clerk Maxwell published "A Treatise on Electricity and Magnetism" which consisted of two volumes (see pages 13 and 14). Toward the end of Volume 2, numerous references are made concerning a "lumeniferous medium" (light carrying medium) through which the light waves (electromagnetic waves) propagate.

Thus, the term "lumeniferous ether" rose to popularity among the leading physicists of the day, as a name for the mysterious medium which (they thought) conducted the electromagnetic waves as they traveled through space. Largely from the information contained in Maxwell's "A Treatise on Electricity and Magnetism", the Ether Theory was developed. This Ether Theory simply stated that there was a "physical medium" (the lumeniferous ether) in which electric and magnetic fields were distortions, and which conducted electromagnetic waves at a rate dependent upon the pressure or density of the medium.

(continued on page 15)

A Treatise on Electricity and Magnetism

Volume 1

James Clark Maxwell

ISBN 0-19-850373-3

A Treatise on Electricity and Magnetism

Volume 2

James Clark Maxwell

ISBN 0-19-850374-1

(continued from page 12)

In the years immediately following the publication of "A Treatise on Electricity and Magnetism", physicists were quite successful in applying this "Ether Theory" in order to explain the "effects" of electricity and magnetism. Further experimentation was conducted, and the predicted electromagnetic waves were successfully transmitted and received in 1887 by Heinrich Hertz (1857 – 1894), opening the door to a whole new field of science and technology.

The early successes of the Ether Theory were not long lived, however. As physicists began to devote more time and effort toward uncovering the secrets of electricity and magnetism, they began to realize that they would need to learn more about the "lumeniferous (light carrying) medium", or "lumeniferous ether". That is when serious problems began to develop for the Ether Theory. One of the most difficult questions was, "How were the electromagnetic waves affected by the Earth's relative motion through space?"

In the 1880's, physicists and astronomers knew that the Earth moves at a rate of approximately 66,000 miles per hour in its orbit around the Sun. They concluded that since the Earth is moving through this "ether" at such a high velocity, there should be a relative motion between the Earth and the "ether" which should bend the light waves slightly.

In 1887, Albert A. Michelson (1852 – 1931), a physics professor at Cleveland, Ohio's Case School of Applied

Science, and Edward W. Morley (1838 – 1923), a chemistry teacher at the Western Reserve University, also in Cleveland, Ohio, began working on a project to prove the existence of this ether. They built a device which would be able to measure to a high degree of accuracy (one part in 100 million) the movement of light waves in different directions with respect to the Earth's movement around the Sun (see page 17).

This is how their experiment worked. They would direct a beam of light toward a thinly silvered mirror which would split the beam, allowing half of the light to pass through, and reflecting the other half towards another mirror. The separate beams of light would then be reflected back and forth in opposite directions before being brought back together again and directed into an eyepiece.

If the light waves arrived at the eyepiece slightly out-of-phase, or blurred, this would indicate that there was a relative motion between the Earth and the "lumeniferous ether". To their great surprise, however, the light waves always arrived back at the eyepiece at the same time, or in-phase. No matter what direction in which they conducted the experiment, the light beams *always* arrived at the eyepiece in-phase. Thus, it became obvious that there was no relative motion between the "lumeniferous ether" in space and the Earth as it moves in its orbit around the Sun.

This experiment has become widely known as the Michelson-Morley experiment. It became famous because it represented an important scientific

(continued on page 18)

THE MICHELSON-MORLEY EXPERIMENT

1887

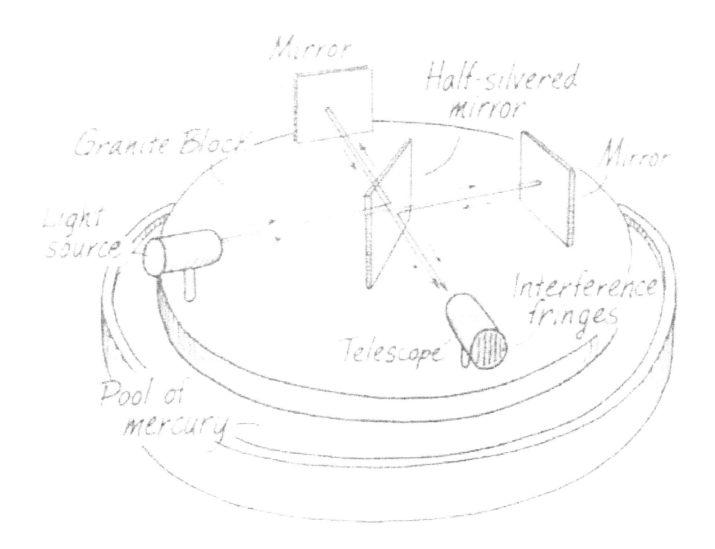

(continued from page 16)

breakthrough which laid the groundwork for modern physics. At first, physicists all over the world were astonished by the results of this experiment. Prior to this experiment, they were almost certain that there had to be some type of relative motion between the Earth and this mysterious "ether" substance which existed out in space.

Another serious contradiction to the Ether Theory which arose shortly afterward was that most of the laws of physics are dynamic, but the laws of electricity and magnetism are not. In order to explain this contradiction, we will need to get a good understanding of what an "inertial reference frame" is.

An inertial reference frame can be compared to an airplane flying at a constant altitude and at a constant velocity above the Earth's surface. Everyone inside the plane can be considered to be in the same inertial reference frame. The people in a different airplane travelling in a different direction, at a different altitude, and at a different velocity relative to the first plane can be considered to be in a different inertial reference frame.

Now consider an astronaut in a spaceship out in space far away from the Earth and travelling at a high velocity relative to the Earth, the inertial reference frame of the spaceship would be considerably different than the inertial reference frame of the Earth or anything else in the universe for that matter. The Earth is travelling in its orbit around the Sun. Our Milky Way galaxy is rotating and flying through space at a high velocity relative to all the other galaxies.

This leads us to the very important question of, "How can we determine which inertial reference frame is absolutely "at rest" or "standing still" within our universe, and which inertial reference frame is moving"? When the astronaut in his spaceship looks out of his window into space, he can only measure his velocity relative to the Earth, or some other celestial body. He has no way of knowing what his true velocity really is, relative to the universe as a whole.

The laws of physics tell us that all inertial reference frames are the same when considering mechanical phenomena within each inertial reference frame, regardless of their relative motion. Then the question arises, How does the speed of electromagnetic waves compare with regard to the relative velocities of the different inertial reference frames? Would the speed of light be greater if the light source were moving toward you at a high velocity, or less if the light source were moving away at a high velocity?

The answer is that the speed of light does not change. Maxwell's equations (see page 20) do not recognize relative velocity, and consequently, they always seek a preferred reference frame --- one that can be considered to be absolutely "at rest" with respect to the universe as a whole. Thus, most of the laws of physics are dynamic, but the laws of electricity and magnetism are static (stationary). This is a very important relationship which reveals the nature of the underlying

(continued on page 21)

The first of Maxwell's equations:	$\oint E \cdot dA = \dfrac{Q}{\varepsilon_o}$
Maxwell's second equation: States that magnetic field lines are continous providing there is no such thing as a magnetic monopole.	$\oint B \cdot dA = O$
Maxwell's third equation: States that an electric field is produced by a changing magnetic field.	$\oint E \cdot dl = -\dfrac{d\Phi_B}{dt}$
Maxwell's fourth equation: States that a magnetic field is produced by an electric current or by a changing electric field.	$\oint B \cdot dl = \mu_o I + \mu_o \varepsilon_o \dfrac{d\Phi_E}{dt}$

(continued from page 19)

physical process which produces the universe that we see. Another surprising contradiction with the Ether Theory is that electromagnetic (light) waves can have both a longitudinal and a transverse polarization. Since transverse waves are possible only in a solid substance, then certain aspects of this "ether medium" must be held in a "fixed state", quite possibly as a result of the laws of physics of the empty void in which the "ether medium" is expanding into. This situation will be described in greater detail in the next chapter.

This, among other circumstances finally led to the demise of the Ether Theory. Tragically, for the Ether Theory, if a particular scientific theory does not describe all of the observations and experimental results which it is supposed to, it is automatically relegated to the trash can and physicists then begin searching for a new theory which will describe all of the results.

For years afterward, many physicists felt that the lumeniferous ether existed in some form. The Maxwell equations, which are still in use today and are widely accepted as being correct, continue to point to its existence. In his famous "A Treatise on Electricity and Magnetism", James Clerk Maxwell wrote: "Hence all these theories lead to the conception of a medium (substance) in which the propagation (of electromagnetic waves) takes place, and if we admit this medium as an hypothesis, I think it ought to occupy a prominent place in our investigations, and that we ought to endeavor to construct a mental representation of all the details of its action, and this has been my constant aim in this treatise".

Perhaps the old Ether Theory was not entirely incorrect after all. Maybe there was just some additional underlying physical process involved which was causing these contradictions and which the physicists were unaware of at that time.

Chapter 2

A Closer Look at Relativity

For several years after the Michelson-Morley experiment was conducted, the scientific community all over the world was in disarray. As a result of the contradictions exposed by the Michelson-Morley experiment and other experiments concerning electricity and magnetism, it was widely recognized that the Ether Theory, and classical physics in general was seriously flawed. For approximately fifteen years, physical science languished without any widely-agreed-upon theoretical basis for the "fixed", or "stationary", effects of electricity and magnetism, or the "dynamics" of length (space), time, relative motion, inertial mass, acceleration (gravity), and energy.

Then in 1905, Albert Einstein (1879 – 1955), who was working as a clerk in the Swiss patent office at the time, introduced a bold new theory which placed more emphasis on how we observe objects and events rather than the strict, mechanistic view of classical physics (see page 25). Einstein determined that the speed of light (electromagnetic waves) remains constant, or "fixed", but that the concepts of time and space are relative (dynamic).

This new theory seemed to reconcile the contradictions of the Ether Theory, while providing a basis for the "fixed" laws of electricity and magnetism as well as the "dynamic" laws of physics. Almost overnight, Albert Einstein became famous within the scientific community. His new theory, which became known as the "Special Theory of Relativity", was studied carefully, and experiments were soon devised to test its accuracy.

With the rising political tensions in Europe and the eventual outbreak of World War I, the accuracy of "The Special Theory of Relativity" and also "The General Theory of Relativity" which Albert Einstein introduced in 1916, was not able to be tested until after the war. Shortly after World War I ended, a British astronomer named Sir Arthur Eddington (1882 – 1944) traveled to an island off the west coast of Africa which would be directly in the path of the solar eclipse of 1919.

(continued on page 26)

Relativity

The Special and the General Theory

Albert Einstein

ISBN 0-517-88441-0

(continued from page 24)

During the eclipse, Sir Arthur Eddington made observations of several stars which, at the time, could be seen just beyond the surface of the sun. The observed locations of the stars were recorded and compared to their previously known locations. To the surprise of many astronomers, The Theory of Relativity accurately predicted that the light (electromagnetic waves) from the stars was bent, or distorted, by the sun's gravitational field. Other tests of The Theory of Relativity were conducted later on, all of which proved its accuracy beyond any doubt.

Classical physics implied that we exist in a three-dimensional (height, length, width) space, upon which we superimpose the concept of time. In Einstein's view, we live in a four-dimensional universe where the co-ordinates of space (height, length, and width) and also time are inseparably linked together and inversely proportional to each other (see page 27. In other words, with an increase in relative velocity, time slows down so that one second lasts longer, or becomes greater, and the three dimensions of space contract, or become smaller, within that particular inertial reference frame.

To produce any type of change in the relative motion between two objects in space, acceleration is required, which in turn, changes the space-time coordinates between the two objects. Hence, for two objects with a difference in space-time coordinates between them, time slows down and length contracts at the rate

(continued on page 28)

LORENTZ TRANSFORMATION EQUATIONS

Graph of the length equation: $\quad L' = L \sqrt{1 - \dfrac{v^2}{c^2}}$

L = Length;   L' = Distorted Length;   v = velocity;   c = speed of light

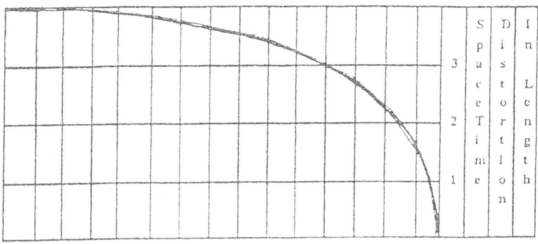

Along the right side, reading vertically: Space Time (3, 2, 1) | Distortion | In Length

← "where we are now"

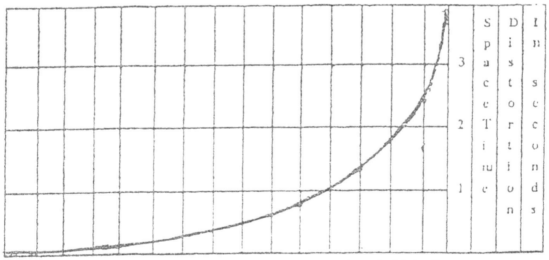

Along the right side, reading vertically: Space Time (3, 2, 1) | Distortion | In seconds

Speed of light " c

"←"where we are now"

Graph of the time equation: $\quad T' = \dfrac{T}{\sqrt{1 - \dfrac{v^2}{c^2}}}$

T = Time;   T' = Distorted Time;   v = velocity;   c = Speed of light

(continued from page 26)

described by the equations on page 27. Change the space-time coordinates around an object in space, and you produce acceleration (gravity).

The Special Theory of Relativity did an excellent job of explaining the latest experimental results while it also eliminated the contradictions of classical physics. Thus, we entered the new era of modern physics in which our observations are determined by the four-dimensional space-time coordinates within our inertial reference frame, relative to other inertial reference frames with different space-time coordinates.

Our entire universe is made up of a large number of different inertial reference frames, each with its own unique set of space-time coordinates. This is what causes us to perceive that the different stars and galaxies lie in a particular direction, at a certain distance, with a certain degree of relative motion, and each possessing its own gravitational field of a particular strength dependent upon its inertial mass (see pages 29 and 30).

In 1929, a few years after The Theory of Relativity had become well established, an American astronomer named Edwin Hubble (1889 – 1953) made another rather startling discovery. Using the powerful 100-inch (254-cm) telescope at the Mount Wilson Observatory in southern California, Edwin Hubble was able to look beyond the stars in our Milky Way galaxy, and observe a large number of distant galaxies:

(continued on page 31)

Unit of Measurement	Relativity Equations	Curve on a Graph
LENGTH This equation describes the rate at which length within an inertial reference frame decreases as relative motion increases.	$L' = L\sqrt{1 - v^2/c^2}$	
TIME This equation describes the rate at which time within an inertial reference frame slows down (1 second becomes longer) as relative motion increases.	$T' = \dfrac{t}{\sqrt{1 - v^2/c^2}}$	
MASS This equation describes the rate at which mass increases within an inertial reference frame as relative motion increases.	$m = \dfrac{m_o}{\sqrt{1 - v^2/c^2}}$	
RELATIVE MOTION This equation describes the rate at which relative motion increases as acceleration is applied during a specified time period.	$v = \dfrac{at}{\sqrt{1 - at^2/c^2}}$	
KINETIC ENERGY This equation describes the rate at which kinetic energy increases in proportion to an increase in relative motion	$v = \sqrt{2KE/m}$	
POTENTIAL ENERGY	$v = \sqrt{2PE/X}$	

GRAVITY This equation desvribes the rate at which an increase in relative motion is caused by the acceleration of a gravitational field which is dependent upon the distance (R) from the center of the object which is producing the gravitational field.	$$v = \sqrt{Gm_e/r}$$	
WAVE MOTION This equation describes the relationship between wave motion and uniform circular motion.	$$v = v_{max}\sqrt{1-(a/x)^2}$$	
$\dfrac{1}{R}^2$ *Law* This term describes the rate at which the acceleration which is produced electric, magnetic, and gravitational fields, decreases with distance (R) from the object which is producing the field. This equation also describes the rate at which the Lumineferous Ether is expanding and spreading out within the empty black void.	$$\frac{1}{R^2}$$	
The Surface of a Sphere This equation describes how the surface of a sphere increases with the distance (R) from its center. This is a fundamental relationship which also represents the expansion of our universe (the expansion of a sphere) .	$$S = 4\pi R^2$$	

(continued from page 28)

As more and more galaxies were discovered, it was determined that virtually all of these galaxies are moving away from us at extremely high velocities with the most distant galaxies moving away at velocities approaching the speed of light. This led astronomers to conclude that we live in an expanding universe.

The cosmological theory which was developed as a result of Edwin Hubble's observations and supported by Einstein's Theory of Relativity, indicates that we are living in a universe which started out as a small, highly concentrated "primordial singularity" approximately 15 billion years ago (as we perceive it). The singularity which consisted of the highly compressed physical substance, has been rapidly expanding and spreading out in all directions ever since like an expanding sphere which now measures approximately 15 billion light years across (as we perceive it).

This theory has become widely known as "The Theory of an Expanding Universe". We are able to conclude that our perceptions of the large area within this large sphere, which we call our universe, are being produced by the four-dimensional space-time energy field which represents the "effects" of the physical substance as it has expanded and spread out. Our perceptions of the dynamics of relative motion, gravity, space (distance), time, acceleration, mass, and energy are all being produced by this expansion process according to the relativity equations and the equations on pages 29 & 30.

Since the effects of relativity are all being produced by the expansion process which our universe is currently undergoing, then the laws of physics of the void which our universe exists within, and is expanding into, must be considerably different than the laws of physics within our universe. These differences are probably the cause of all the contradictions regarding the old ether theory and the theory of relativity, and also the discrepancies between the theory of relativity and quantum physics.

But what about electricity and magnetism? One of the most prominent of the mysteries is that the laws of electricity and magnetism do not obey the relativity principle. Many experiments which were conducted with electric and magnetic fields and also electromagnetic waves, seem to choose a superior reference frame which is always "at rest" with regard to the relative motions of the Earth and all of the other stars and galaxies.

Shortly after this discovery was made, it was brought to Albert Einstein's attention that the time and length (Lorentz transformation) equations in his Special Theory of Relativity actually describe this expansion process, but in the reverse direction (see the graphs on page 27). When these two equations are solved for greater and greater relative velocities and then plotted on a graph, they describe a pressure release of a highly pressurized physical substance into empty space.

The inverse relationship of these two equations is showing us that as the size of the spherical cloud of this physical substance is becoming larger, the pressure, or density, of it is going down in direct proportion. When taken at face value, these two equations are describing a very important, underlying physical process which is actually producing our perceptions regarding the effects of relativity. Even more surprisingly, it becomes apparent that the process which these two equations are describing is actually what is causing time to pass as we know it.

Judging from the time and length equations, the effects of relativity which are producing our perceptions of length (space), time, relative motion, inertial mass, acceleration (gravity), and energy, are all being caused by distortions in the pressure-related expansion process of this mysterious physical substance. Could this physical substance possibly be the mysterious and elusive lumeniferous ether?

As a result of this new evidence, we can now reexamine the results of the Michelson-Morley experiment in a whole new perspective. When considering the relative motions of the Earth, it now becomes obvious that the Earth is not really moving through the physical substance, per se. But instead, distortions in the four-dimensional space-time energy field, which was produced by the expansion of the physical substance is what is actually producing our perceptions about the relative motions of the Earth.

We can now see that the way the Michelson-Morley experiment was set up, it is no wonder that it failed to detect the presence of the substance itself. Since there was no relative motion between the light source and the detector (the whole unit was mounted on a solid platform), there could not possibly have been any phase shift between the two light beams which were being compared by the detector regardless of which direction the experiment was being conducted.

This was simply the wrong type of experiment for detecting the actual physical presence of the substance. In hindsight, we can also see how ridiculous it was for these physicists to become so upset when an experiment does not produce the desired results. Often, we can learn as much from an experiment which fails to produce the desired results as we do from one that does.

At the introduction of the Theory of Relativity, these early physicists were proven wrong about the way in which the real physical properties of the substance (lumeniferous ether) are manifested in our time frame, but not necessarily wrong about the existence of the substance itself. The mysterious and elusive lumeniferous ether just might be the missing link which would help bridge the gap between the "fixed" effects of electricity and magnetism and the "dynamic" effects of relativity.

According to Albert Einstein's General Theory of Relativity, gravity is caused by the same type of distortion in the pressure-related expansion process which produces our perceptions of relative motion and makes up our four-dimensional space-time continuum in general. If electromagnetic waves can be distorted by entering a gravitational field (gravitational red shift), then there has to be some common denominator in this mysterious underlying physical expansion process which is creating our perceptions of time and space, and also our perceptions of electricity and magnetism.

The Relativity Theory, however, still left several

troublesome questions unexplained. Somehow, the effects of gravity, electricity, and magnetism must all be related to the universal common denominator which lies just beyond our reach. A strange set of circumstances arises in which you have the lumeniferous ether which remains in a "fixed" state on one hand. On the other hand, you have the dynamic (changing) effects of relativity resulting from a physical expansion process.

There seems to be no way possible to reconcile these two different sets of circumstances. However, in each case which we have studied so far, when we begin to encounter contradictions and unexplained circumstances with an existing theory, this is a sure indication that there is yet another hidden, underlying, physical process which is going on behind the scene to produce these contradictions.

Chapter 3

THE "EMPTY BLACK VOID"

Up until the early 1930's, most physicists believed that we live in a "steady state" universe largely because much of the scientific research which had been conducted during the 1800's and early 1900's had ever given them any reasons to believe otherwise. Maxwell's equations, which were widely accepted and well established by 1880, always seemed to demand a preferred reference frame which is always "at rest" with respect to our universe as a whole.

It was also widely accepted that electric and magnetic fields were a property of free space, which gave physicists the impression that at least some aspects of free space always remain in a "fixed state". Furthermore, since many of the values which are used in the equations associated with electricity and magnetism are constants, Albert Einstein never even suspected that our universe might be expanding, even while he was writing down the very equations which described this expansion process, in his Special Theory of Relativity.

This situation leads us to make some very interesting observations. First, we have the classical physicists who have devoted their entire lives to studying electric and magnetic phenomenon, believing that there is a mysterious physical substance which is somehow responsible for transmitting the effects of electricity and magnetism through space.

Secondly, they believed that this mysterious physical substance which is associated with electric and magnetic phenomena remains in a "constant", or "fixed" state, and thus, our entire universe should remain in a "fixed" state. This set the stage for Edwin Hubble's surprise discovery that we do, in fact, live in an expanding universe. The nature of this mysterious and elusive lumeniferous ether might just be the missing link which would bridge the gap between the "fixed" effects of electricity and magnetism and the "dynamic" effects of relativity.

The "Big Bang Theory" which was derived from Edwin Hubble's observations and The Theory of Relativity, suggests that our universe started out as a small, highly concentrated, "primordial singularity" approximately 15 billion years ago. The time and length equations in The Theory of Relativity indicate that this singularity consisted of a highly compressed physical substance, which suddenly began expanding and spreading out, very rapidly at first. Then, as the size of the primordial singularity got bigger, the pressure of the physical substance of which the primordial singularity consists, began to decrease, in direct proportion, as the size of it increased.

This singularity has been rapidly expanding and spreading out in all directions ever since, like an expanding sphere which now measures approximately 30 billion light years across (as we perceive it). Graphs of the Lorentz transformation equations, which were used in Albert Einstein's Special Theory of Relativity, describe how this physical expansion process took place (see page 27). With this type of evidence, we are now able to determine that there must have been some type of physical substance involved which is driving the universal expansion process which we are observing.

The time and length equations, taken together, demonstrate that as the size of the spherical cloud of this substance increased, the pressure of it decreased proportionately. Thus, we have the evidence of a mysterious physical substance being involved in the basic, underlying, physical process which formed our universe. Ironically, the strongest evidence that we have which indicates that this "physical substance" really exists, comes from The Theory of Relativity which was supposed to replace the Ether Theory.

The time and length equations describe this "pressure-related" expansion process when they are solved for greater and greater relative velocities and then plotted on a graph. The graphs of these equations provide us with compelling evidence that our early universe probably began as a very small singularity (10^{-15} cm) in diameter, then, as it grew larger, it became an expanding sphere which has continued to expand right up until now.

Even now, it continues to expand, and from all indications, this continuing expansion process is what is causing time to pass as we know it. If this basic description of our universe withstands the tests of further experimentation and observations, another set of very interesting questions will then be raised.

A) What is the nature of the void in which our universe came into existence and then began expanding into?

B) How are the laws of physics in this void different from those within our universe?

C) How does the nature of this void interact with the mysterious "physical substance" to produce the effects of relativity and also the effects of electricity and magnetism?

These questions may seem rather difficult when we first encounter them. Of course, this can become a very technical subject. However, as the novice becomes more familiar with the terminology and the different ways of proving these physical relationships, this new field of science actually becomes quite familiar and easy to understand.

There is a considerable amount of experimental evidence available to demonstrate how these physical relationships interact to produce the perceptions we have about the world we see around us and our universe in general. Perhaps the most important of these perceptions is that the speed of light remains constant while time and space are relative.

The first step toward an explanation of this relationship is the realization that what we consider to be "empty space" within our observable universe is not really empty. If we could travel out into intergalactic space, our first observation would be the beautiful array of all different sizes and shapes of the rather large number of galaxies which would be close enough for us to observe.

This emphasizes the fact that there is an enormous amount of electromagnetic wave energy at all different frequencies traversing every square inch of the space within our universe every second. When you think about it, all of the tremendous amount of electromagnetic wave energy being released every second from all of the stars in all of the galaxies is actually being stored in the "space" within our universe while it is travelling from one place to another.

Gravitational fields tug and pull at each other across vast reaches of intergalactic space. In addition to this, our observations have shown that there are also small amounts of hydrogen and helium atoms existing in intergalactic space. So what we consider to be empty space within our universe is certainly not empty, it is actually a very busy place.

Now that we have established the fact that all of the space within the huge sphere which we call our universe is filled and occupied with something, even though it may not be readily visible to the observer, we need to consider the space which lies outside the outer reaches of our universe --- the void which our universe is expanding into. Since the relativity equations tell us that our perceptions of time and space are being produced by the expansion process which is taking place within our universe, then outside our universe the concepts of time and space, as we know them, would be invalid or meaningless.

In this case, the void outside our universe would be completely empty space, or "dead space". There would be no four-dimensional space-time energy field, which is represented by the Permittivity and Permeability values, to conduct the gravitational and electromagnetic fields, and produce our perceptions of time and space. In other words, time, as we know it, would be completely stopped and length, or size, would be contracted down to 10^{-15} cm which is comparable to a single point.

Now this is precisely what The Theory of Relativity tells us about two objects which have a high degree of relative motion, time slows down and length contracts in regard to each of the objects. This relationship demonstrates how the laws of physics in the empty void are able to influence our laws of physics and other events within our universe.

If we live in an expanding and totally dynamic universe according to our observations, then why are there any constants like the speed of light, the gravitational constant, Boltzmann's constant, the Permittivity "constant", the Permeability "constant", Planck's constant, etc. Why would anything in our universe remain in a "fixed" state unless there are two completely different sets of "laws of physics" which govern the space within our universe and the space which lies outside the outer reaches of our universe.

The physical substance which is driving the expansion of our universe is "remembered", or "held" by the void in its original, highly pressurized state. This gives the physical substance a dual existence --- extremely high physical pressure, concentrated within the singularity when it was in its original state, on one hand, and also the greatly reduced pressure (almost to zero) as it has expanded and spread out, on the other hand, simultaneously. Consequently, this tremendous, simultaneous pressure difference is what produces the four-dimensional space-time energy field which, in turn, produces our effects of relativity.

The term "four-dimensional space-time energy field" is used to describe this abstract energy field which produces our effects of relativity. The four-dimensional space-time energy field is considered to be "abstract" because it has no real physical presence in the space within our universe, it only represents the "effects" of the real "physical substance". The main reason that it is referred to as a four-dimensional space-time energy field is because that's what is causing us to perceive that our universe has expanded to its present size (space) and that roughly 15 billion years have passed since it began to expand (time).

The effects of relativity are what produces our perceptions of length (space), time, relative motion, inertial mass, acceleration (gravity), and energy. This set of circumstances leads us to conclude that the "physical substance" has a dual existence. It appears to remain in a "fixed" state (within the singularity) with respect to electricity and magnetism, while the relativity equations tell us that is continuously changing (dynamic) by expanding and spreading out.

The more you think about it, the more you must realize that this void in which our universe exists would certainly seem to us like a very strange and unusual place. Never in the history of mankind have we ever encountered anything like this. First of all, from The Theory of Relativity, we have learned that no massive object can ever reach or exceed the speed of light in order to reach the outer edge of our universe, so we will probably never be able to go there and study it directly.

Secondly, we all depend on the four-dimensional space-time energy field, which is represented by the Permittivity and Permeability values (see page 123), for our very existence. The atoms which we are made of would disintegrate if we were able to travel there. The strange nature of this void might also be causing mathematical problems for the physicists and mathematicians who are seeking to discover a Grand Unified Theory solely by the use of mathematics.

There is an enormous amount of time and effort being spent by many physicists and mathematicians in colleges and universities around the world in the search for a complete mathematical description of our universe. Among the most noteworthy of these physicists is Professor Stephen W. Hawking (1942 – 2018) who held the prestigious "Lucasian Chair of Mathematics" at Cambridge University in England --- a position once held by Isaac Newton.

Many of our leading physicists use mathematics to advance our knowledge in physics because mathematics has served us so well in the past. However, in the past few years, we have been running into situations which are quite difficult to be fully explained by mathematics. By relying on mathematics as their only source of new information, physicists are probably not going to get all of the information which they will need to formulate a complete Grand Unified Theory.

Mathematics is a precise and effective method of communicating four-dimensional space-time distortions gravitational fields), electric fields, magnetic fields, probabilities, numerical values, and virtually everything we see going on around us. But what about the void? Do we have a system of mathematics which will completely describe the laws of physics in this void, and describe the strange relationship between the "physical substance" and this void in which our universe exists? Of course we do not have the answer to this, and that is the main reason that we have not been able to link The Theory of Relativity and Quantum Theory mathematically.

In the next chapter, you will see how we can learn more about the nature of this void from the results of quantum experiments. By analyzing the results of a large number of experiments and arranging the information into different categories, we will be able to learn more about the underlying physical process which is producing the universe that we see. Also, we will be taking a much closer look at electric and magnetic fields and their close relationship to the effects of relativity which include mass, energy, gravity, and relative motion.

Chapter 4

ELECTRIC AND MAGNETIC FIELDS

Have you ever held two magnets together and wondered what strange force pulls them together or holds them apart? When you use two large and powerful magnets, it becomes quite obvious that these magnets are able to muster a considerable amount of energy as they attract or repel each other. The invisible force acting between the magnets must be a property of the space around the magnets.

Now, let us consider an electromagnetic wave travelling

through intergalactic space, far away from any massive objects or strong gravitational fields. In a region of space such as this, there are very few atoms or gravitational fields to interfere with an electromagnetic wave as it passes by. As the portion of the wave which contains the negative electric charge enters this region of free space, the free space is charged to a negative electric potential without any electrons being present.

In other words, what we consider to be empty space within our universe has the ability to become negatively charged as a property of space itself. The same holds true for the positive electric charge and corresponding north and south magnetic fields. These fields do not depend on the presence of electrons or iron atoms or anything else in outer space because they are simply a property of the space within our universe (Permittivity and Permeability of free space).

The Permittivity of free space represents the amount of capacitance, or the ability of free space to contain an electric charge. The Permeability of free space represents the inductance, or the ability of free space to contain a magnetic field (see page 123). The fact that free space has the ability to contain an electric charge and a magnetic field should come as no surprise to most physicists.

Ever since James Clerk Maxwell's time, physicists have used the Permittivity and Permeability of free space to calculate the speed at which electromagnetic waves

(continued on page 51)

MAGNETIC FIELDS

INDUCED CURRENT

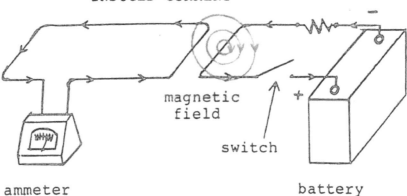

magnetic field

switch

ammeter

battery

The current flow which is shown here, represents the <u>actual</u> movement of the electrons through the wire. The <u>apparent</u> current flow would be in the opposite direction.

An electron represents an area of decreased physical "pressure" in four-dimensional space-time (as it exists at the point of "basic reality"). It is held in that condition because of the effects of relativity on its rapidly oscillating negative half-wave-form.

When the switch is closed, the electrons in the wire connected to the battery are placed in motion (accelerated) which creates velocity-related space-time distortion within the electrons along their axis of movement. This causes the four-dimensional space-time itself (which has a "real" existence only at the point of "basic reality") in the vicinity of the electrons to undergo acceleration at a right angle to the direction in which the electrons are moving.

The resulting acceleration of the four-dimensional space-time around the wire (magnetic field), in turn, accelerates the electrons in the nearby wire at a right angle to the magnetic field, but in the opposite direction than the electrons which are moving in the wire connected to the battery.

HYDROGEN ATOM

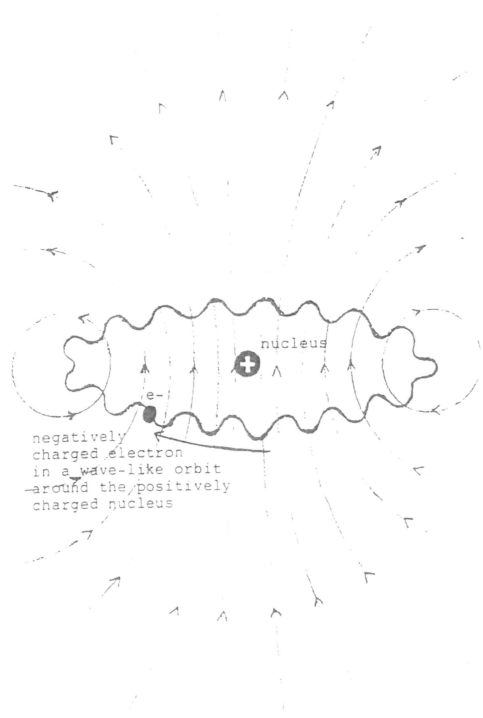

nucleus

e-

negatively
charged electron
in a wave-like orbit
around the positively
charged nucleus

these magnetic lines of flux represent the
movement of four-dimensional space-time
created by the moving electrical charge
(electron).

BAR MAGNET

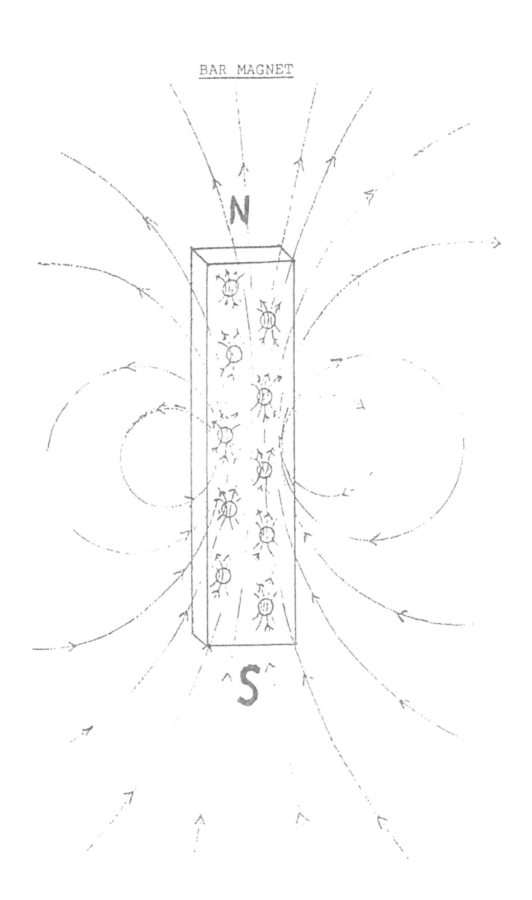

(continued from page 47)

travel, or propagate, through space, which is commonly referred to as the speed of light (see page 123). In intergalactic space, or relatively empty space within our universe, the Permittivity and Permeability values tend to remain fairly constant with the exception of the small vibrations corresponding to the passing electromagnetic wave energy (light) and minor gravitational attractions between neighboring galaxies.

Thus, we have an explanation of why the speed of light tends to remain fairly constant, in most cases, while time and space are relative. It is this capacitance and inductance of free space which stores, or contains, the electric and magnetic fields of the electromagnetic waves briefly as they pass through.

This is how free space is able to store such an enormous amount of electromagnetic wave energy (light, radio frequency waves, microwaves, X-rays, gamma rays, cosmic rays, etc.) which is being produced by all the stars in all the galaxies, from the time that our early universe first started producing light, until today. Of particular consequence to us, is the storage and conduction of all the electromagnetic wave energy coming from the sun during the 8.5 minutes while it is travelling to the Earth.

This relationship also indicates to us that the universal four-dimensional space-time energy field, which is represented by the Permittivity and Permeability values, is not only producing our perceptions of electric and magnetic fields, but also our perceptions of time and space (gravitational fields, relative motion, mass and energy), as well. This leads us to conclude that the Permittivity and Permeability of free space, are not really constants after all, and so, under certain circumstances, they can become variable.

A positive electrical charge, then, may be considered to be a high-pressure distortion in the "physical substance" which is represented by an increase in the Permittivity value by itself (see page 53). A negative electrical charge can be considered as a low-pressure distortion in the "physical substance" which is represented by a decrease in the Permittivity value by itself for a particular region of space (see page 54).

Our perceptions of positive and negative electric charges are a direct result of the specific Permittivity value at all points in space where the positive or negative electric charges are located. In addition to this, physical pressure differences (positive and negative electric charges) in the actual "physical substance" are always perpendicular to expansion-related (velocity-related) distortions such as: gravity, relative motion, space, time, mass, and energy; in the "physical substance".

The expansion-related (velocity-related) distortions in the four-dimensional space-time energy field are a direct result of "equal" variations in both the Permittivity and Permeability values. Consequently, the difference in the Permittivity values in the space surrounding electric charges are responsible for the attraction (acceleration) of two opposite electric charges, and the repulsion

(continued on page 58)

POSITIVE ELECTRIC CHARGE

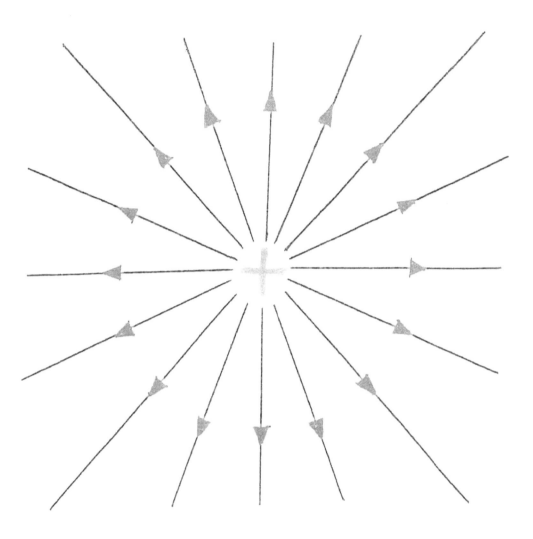

This drawing describes a region of space which is under the influence of a higher "pressure" (than the median "pressure") of the "physical substance" within the primordial singularity as it exists in the realm of "basic reality".

NEGATIVE ELECTRIC CHARGE

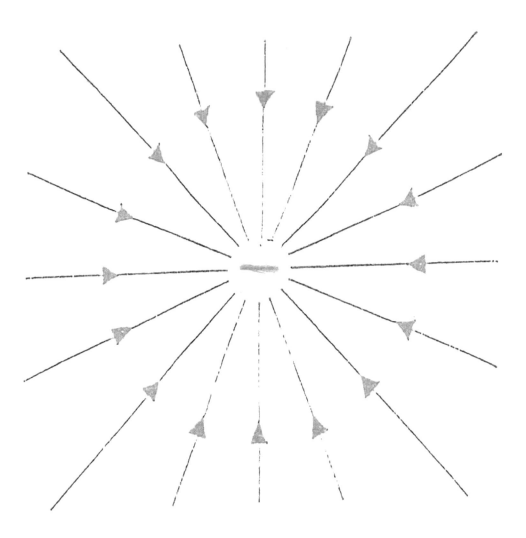

This drawing describes a region of space which is under the influence of a lower "pressure" (than the median "pressure") of the physical substance within the primordial singularity as it exists in the realm of "basic reality".

TWO POSITIVE ELECTRIC CHARGES

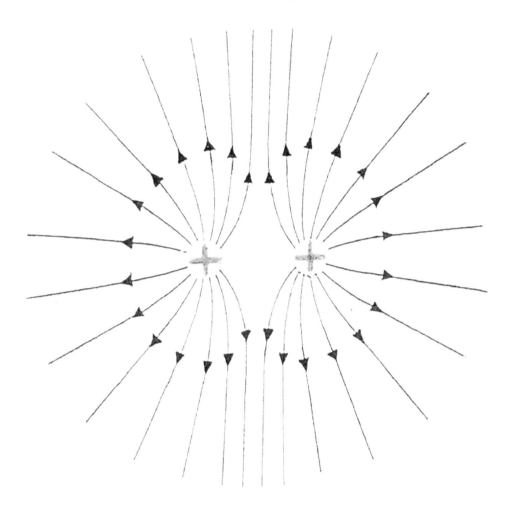

The two positive electrical charges shown here represent regions of space which are under the influence of a higher "pressure" (than the median "pressure") of the "physical substance" within the primordial singularity. Since physical "pressure" differences in the "physical substance" always tend to seek the median "pressure" or equilibrium, as it exists at the point of "basic reality", then these two charges will repel each other in order to remain as close to equilibrium as possible.

TWO NEGATIVE ELECTRIC CHARGES

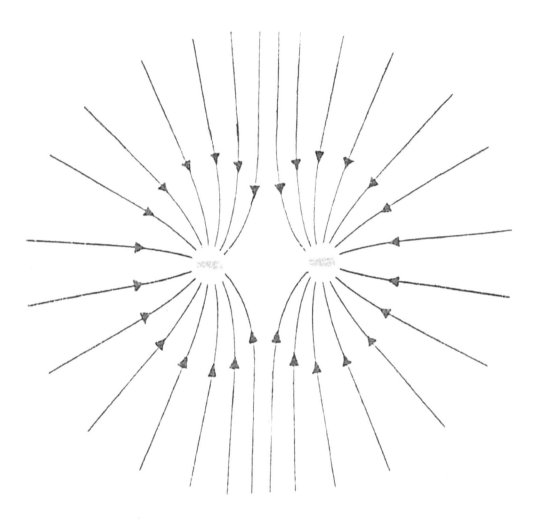

The two negative electrical charges shown here represent regions of space which are under the influence of a lower "pressure" (than the median "pressure") of the "physical substance" within the primordial singularity, Since physical "pressure" differences in the "physical substance" always tend to seek the median "pressure" or equilibrium, as it exists at the point of "basic reality", then these two charges will repel each other in order to remain as close to equilibrium as possible.

OPPOSITE ELECTRIC CHARGES ATTRACT

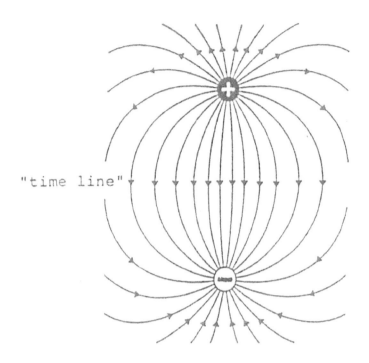

"time line"

Opposite electrical charges tend to attract because they represent "pressure" differences in four-dimensional space-time. This is why they are shown on opposite sides of the "time line", which is represented by the horizontal line of arrows above.

The acceleration that they produce is always toward equilibrium, or the median "pressure" of four-dimensional space-time as it exists at the point of "basic reality". The point of "basic reality" (in this case) represents the "real" existence of the four-dimensional space-time in our universe as we see it (our dimension).

58

(continued from page 52)

(acceleration) of two like electrical charges. Also, the Permittivity and Permeability values are perpendicular to each other which has been demonstrated by experimentation regarding electromagnetic phenomena.

The acceleration between electric charges and magnets is the same type of acceleration you feel in a gravitational field, and in all three cases they are governed by the $\frac{1}{R^2}$ law which demonstrates that they are all being produced by the same underlying physical process.

In 1919, Sir Arthur Eddington traveled to an Island off the west coast of Africa, which would be directly in the path of an upcoming solar eclipse. During the eclipse, he conducted observations of several stars, which, at that particular time, could be seen just beyond the surface of the Sun.

The observed locations of the stars were recorded and compared with their known locations later. Sure enough, our Sun's gravitational field had bent or "lensed" the light coming from the stars by the amount predicted in Einstein's theory of relativity. This proves that our four-dimensional space-time continuum is distorted in a gravitational field the same way it is distorted by acceleration to produce our "perceptions" of relative motion.

In the modified time and length equations on page 60, you can see that all of the key values: acceleration; time; length; the Permittivity and Permeability values; are all connected in one set of equations. By mentally comparing these relationships, you can now visualize what will happen if the Permittivity and Permeability values (ε_0 and μ_0) do change, say, in a gravitational field, for instance.

When ε_0 and μ_0 are changed, or altered, in a gravitational field, --- that causes T' and L' to change also. Consequently, it is the T' and L' changes (space-time distortion) which produces the acceleration that you feel in a gravitational field in addition to the electromagnetic effects which include "gravitational redshift" and the "lensing" of the light waves that was observed by Sir Arthur Eddington in 1919.

An experiment which needs to be conducted in order to better-understand the Permittivity and Permeability values, and how they are producing the "effects" of our four dimensional space-time continuum, would be to develop a more broad-based measurement system for these two values so that all the different aspects of them can be easily measured in just about any particular region of space. Secondly, we should accurately measure these values on a high mountain, at around 12,000 feet above sea level, where the Earth's gravitational field is weaker, and then compare these measurements with the values which were derived near the Earth's surface.

We need to learn more about how electromagnetic field

(continued on page 61)

TIME AND LENGTH EQUATIONS IN THE SPECIAL THEORY OF RELATIVITY

For the "time" equation (see page 15) we have:

$$T' = \frac{T}{\sqrt{1 - \frac{v^2}{c^2}}} \; , \text{ Then substitute } (a \times T)^2 \text{ for } v^2$$

Since $c = \frac{1}{\sqrt{\varepsilon_0 \mu_0}}$, you can substitute $\frac{1^2}{\varepsilon_0 \mu_0}$ for c^2

This gives: $T' = \dfrac{T}{\sqrt{1 - \dfrac{(a\,T)^2}{\varepsilon_0 \mu_0}}}$ for the "time" equation.

Likewise, for the length equation   we have:

$$L' = L \sqrt{1 - \frac{v^2}{c^2}} \; , \text{ Then substitute } (a \times T)^2 \text{ for } v^2$$

Since $c = \frac{1}{\sqrt{\varepsilon_0 \mu_0}}$, you can substitute $\frac{1^2}{\varepsilon_0 \mu_0}$ for c^2

This gives: $L' = L \sqrt{1 - \dfrac{(a\,T)^2}{\varepsilon_0 \mu_0}}$ for the "length" equation.

T' = distorted time

L' = distorted length

T = time in "flat" space (away from any strong gravitational fields).

L = length in "flat" space (away from any strong gravitational fields).

c = speed of light in a vacuum

a = acceleration

v = velocity

T = time in "flat" space (away from any strong gravitational fields).

ε_0 = Permittivity of free space: 8.854,187,8E-12 Farads per meter

μ_0 = Permeability of free space: 1.256,637,061,44E-7 Henry's per meter^{-1} (The $^{-1}$ exponent means that μ_0 is inversely related to ε_0)

(continued from page 59)

oscillations are slowed down, or "stretched out" in a gravitational field (gravitational red shift), and how the waves are directionally shifted to produce the "lensing" of the light waves. The amplitude (electric and magnetic field strengths) of the electromagnetic waves, as well as the speed at which the waves propagate, do not seem to be effected very much by the gravitational field.

This is due to the fact that the amplitude of the electromagnetic waves, and the rate at which the waves travel, represent the actual physical characteristics of the electric and magnetic fields, which have a real existence only within the primordial singularity which is held in a "fixed" state by the nature of the "empty black void".

The proof of this statement can be found in the fact that electromagnetic waves come in two basic types: The first type is a **transverse wave**, which can be compared to the up and down motion of a wave on a tight string. This type of wave motion is always perpendicular to the direction of travel, or propagation of the wave. The second type is known as a **longitudinal wave**, or compression wave in which the compressions and de-compressions of the medium in which the wave exists, lie along the same axis as the propagation of the wave.

Even though electromagnetic waves can have both, a transverse and a longitudinal polarization, the transverse wave polarization of electromagnetic waves are what we find most interesting because laboratory experiments have shown that transverse waves are able to propagate only in a solid.

Laboratory experiments performed many years ago have shown that longitudinal waves can propagate through both, a solid or a liquid, but that transverse waves can only propagate through a solid because there has to be a "restoring (holding) force" in the medium to enable them to propagate. In other words, the conducting medium has to be held stationary, or in a "fixed state", in order to enable the transverse waves to propagate.

This evidence was used many years ago to determine that the core of the Earth is molten. During an earthquake both longitudinal ("P" waves) and transverse ("S" waves) are produced and can be detected because both types of waves can travel through the Earth's solid crust. However, due to the fact that there is no restoring (holding) force in the liquid (molten) core, the longitudinal waves were able to travel right on through the core and could be detected on the opposite side of the Earth, but the transverse waves did not.

This evidence makes a rather convincing argument that the "physical substance" in the primordial singularity is being held in a "fixed state" by the laws of physics of the void in which it exists (realm of "basic reality"). Consequently, our perceptions about electromagnetic waves are being produced by a combination of "pressure" and "change" types of distortions in this "physical substance" as it is being held "fixed" or stationary.

The fact that Albert Einstein presented this evidence in one (perhaps even more) of his lectures (see page 105) makes this whole situation truly remarkable because the "theory of relativity" is what was *supposed* to have made the previous "ether theory" obsolete. Surprisingly, instead of disproving it, the theory of relativity provides us with some of the most convincing evidence which *supports* it in a somewhat modified form.

This evidence includes the mathematical description of the physical expansion process by the "time" and "length" equations (see page 27), and here, we are seeing even more convincing scientific evidence that the "physical substance" is being held in a "fixed" state by the nature of the void in which it exists. This is also why we have "constants" in physics.

In the next chapter, you will see how we can learn more about the nature of the "empty black void" from the results of quantum experiments. By analyzing these results, and then by arranging the information into different categories, we will be able to learn more about the underlying physical process which is producing the universe that we see. Also, we will be taking a much closer look at electric and magnetic fields and their relationship to the effects of relativity.

Chapter 5

FOUR-DIMENSIONAL SPACE-TIME

During the early 1930's, the fact that the Theory of Relativity contained evidence of an expanding universe was brought to the attention of Dr. Albert Einstein. At that time, he realized that he had made a serious mistake in regard to his use of a "cosmological constant" which he had included in his General Theory of Relativity which he introduced in 1916. Dr. Albert Einstein had used this "cosmological constant" to mathematically represent a "fixed", or, "steady state" universe because the physical constants and many other aspects of electricity and magnetism appear to choose a preferred reference frame which is always "fixed", or, "at rest" in comparison to all other reference frames, when we actually live in an expanding (dynamic) universe according to Edwin Hubble's observations.

Ever since Edwin Hubble's surprise discovery that we live in a constantly expanding, or dynamic universe, the question then arose, "Why do we have all these physical constants which tend to remain in a "steady", or, "fixed" state (see page 124) while our universe "appears" to astronomers as if it is continually expanding and spreading out?" What mysterious, underlying physical process is at work which would hold some things "fixed" while our universe "appears" to us to be expanding and spreading out?

In physics, we encounter this contradiction over and over. Certain aspects of electricity and magnetism appear to remain in a "fixed" state, including Planck's constant, the capacitance of free space (ε_0), the inductance of free space (μ_0), the speed of light, the charge on an electron, and the charge on a proton, to name a few (see page 124), while the relativity equations demonstrate that the "physical substance" is constantly changing --- expanding and spreading out.

In the previous chapter, we determined that we need to categorize our experimental evidence in a logical sequence in order to formulate a unified theory which will reveal the underlying physical process which is producing some of the "fixed" effects of electricity and magnetism, and also the effects of relativity. Since most of the experimental evidence which has been discovered, has already been fitted into a major theory, these major theories are arguably the best place under which to categorize our latest experimental results.

Once our information has been divided and categorized among the major theories, then all we have to do is devise a new theory which encompasses all this information, and then narrow it down to a single underlying physical process. Almost all of our experimentally discovered information and observations can be categorized under four main theories: The Ether Theory; The Theory of Relativity; the Theory of an Expanding Universe; and Quantum Theory.

Until now, we have these four different theories, each with a considerable amount of experimental and observational evidence to back them up. But each of these theories is not entirely compatible with the others. This strongly suggests that there is still some hidden, underlying physical process taking place which the physicists were not aware of when each of these theories was being developed.

Due to the fact that we already possess a very large amount of experimental and observational evidence in every major area of physical science, all we need to do is arrange the information that we already have under these four basic categories. These categories are:

1. The Ether Theory: Describes the results of some of the early experiments which were conducted to learn more about the effects of electric and magnetic fields. This theory suggested that the electric charge was a pressure distortion, and that a magnetic field is a type of "change" distortion in a mysterious physical substance which was named "lumeniferous ether" and which supposedly filled all the space within our universe, and served as a conducting medium for electromagnetic (light) waves.

2. <u>The Theory of Relativity:</u> This theory does a magnificent job of explaining the experimental and observational evidence regarding the "real" pressure-related expansion of our universe as it continues to expand and spread out within the "empty black void". This theory describes much of the experimental and observational evidence concerning relative motion, gravity, potential and kinetic energy, inertial mass, acceleration, and also the concepts of time and space (as we understand them).

3. <u>The Theory of an expanding universe:</u> This theory describes the expansion of our universe from a cosmological standpoint which suggests that our universe started out as a small, highly compressed, primordial singularity approximately 15 billion years ago (as we see it), and has been expanding and spreading out in roughly a spherical pattern ever since.

4. <u>Quantum Theory:</u> This theory describes the experimental evidence concerning the dual personality of electromagnetic wave/particles, which are actually "fixed" distortions within the "primordial singularity", and which are influenced by the effects of relativity. This theory also contains the experimental evidence which describes the inter-relationship between the physical substance, within the "primordial singularity", and the "empty black void".

These four theories provide us with the basic relationships which govern everything in our universe. First we have the situation where the physical substance remains in a "fixed" state within the "empty black void". Secondly, we have the scenario in which the physical substance is continuously expanding and spreading out into the "empty black void" and producing the effects of relativity. The third relationship concerns electromagnetic waves which exist as "fixed" distortions within the primordial singularity, and, consequently, are influenced by the effects of relativity (the four-dimensional space-time energy field, also represented by the permittivity and permeability constants) to produce our perceptions of energy and mass.

Practically every physical process within our entire universe can be considered to be the result of a combination of these basic relationships. These are the fundamental principles which describe practically all of the physical processes which are occurring all around us. The solution to our "Unified Theory" challenge is probably going to be related to the strange nature of the "empty black void" because that is the most likely "common denominator" which has not been mentioned previously in any of these other theories.

If there is going to be a "common denominator" which unifies these theories, it is probably going to have something to do with the nature of this "empty black void".

Until now, we have seen how the expansion of the "physical substance" into the "empty black void" has created our universal four-dimensional space-time energy field which, in turn, produces our perceptions of space and time as well as our perceptions of electric and magnetic fields which also includes electromagnetic (light) waves. Even so, this hypothesis still raises additional questions which indicate that the laws of physics, in the realm outside the outer reaches of our universe, are very much different, and also have a very profound influence on events within our universe.

Earlier, we came to the conclusion that the "physical substance" started out as a highly pressurized primordial singularity within the "empty black void", then, as the size of the "cloud" of this substance became larger, the pressure of it went down proportionately. Since the concepts of time and space, as we understand them, are meaningless within the "empty black void", the only way that the primordial singularity could expand and spread out according to the graphs on page 27, was for the "physical substance" to take on a whole new existence each time the primordial singularity got a little bit larger.

Each separate existence of the primordial singularity would then represent a different size and pressure of the "cloud", or, "sphere" of the "physical substance" of which it consists (see page 118). This process would have the ability to produce a significant number, perhaps 15 – 20 separate universes similar to our own, providing that the "physical substance" in each separate singularity was small enough and sufficiently concentrated enough to produce good strong neutrons, which of course, would then decay and produce atoms.

The overall expansion process would then be producing the dynamic effects of relativity within each separate universe. Consequently, the four-dimensional space-time energy field which is representative of the size changes, and pressure changes, of the "physical substance", within all of the separate existences, would also be influenced by all the mass within each of these separate universes.

These separate universes which we are unable to observe, but are all interconnected by the same physical substance, are probably where all the "dark matter", and "dark energy" is located because it would still have an overall effect on the four-dimensional space-time energy field which is represented by the Permittivity and Permeability values in our existence (or dimension).

Since protons, neutrons, electrons, photons, and other subatomic electromagnetic wave/particles are so very small within our universe, then the particular existence of the primordial singularity which produces the universe that we see, must have come into existence very early in the expansion process while the primordial singularity was still very small and highly concentrated. We know this because physicists have been able to determine the physical size (diameter) of a neutron, which is representative of the primordial singularity in our universe ("Dimension"), to a relatively high degree of accuracy (1.65×10^{-13} cm).

If you think about this process, without the properties of space and time, there can be no relative motion, wave motion, gravity, mass, energy, or anything else, as a result, this mysterious "physical substance" has to possess several very unusual characteristics. The first of which is its ability to expand and spread out within the "empty black void". Considering the laws of physics in the "empty black void", it is surprising to discover that this expansion process was ever able to take place at all.

From a technical standpoint, the "physical substance", which the primordial singularity consists of, could have remained in its original "fixed" state, forever, without ever having expanded and spread out into the "empty black void". From the relativity equations, however, we do know that, by taking on different existences, the "physical substance" was able to expand and spread out into the void.

From The Special Theory of Relativity we learn that the point of view from which we observe events within our universe is all-important as we formulate our opinions about what we think we see. This point of view consists of a specific set of space-time coordinates called an inertial reference frame which is different for each observer.

The reason for this is, each observer has a different location in four-dimensional space-time (there is some space, or distance, between them) and there may be gravitational or relative motion differences between them also. So each observer, in his own reference frame (point of view), has his own opinion about what he sees around him.

Those different observers may disagree about the distance to, or the size of a certain object. They may disagree about the rate at which time is passing, depending upon their relative motion, or whether or not they are in a strong or weak gravitational field. Direction of relative motion is another factor which helps to determine the overall space-time coordinates (reference frame) for each observer.

One of the most important postulates of The Theory of Relativity is that the speed of light, the speed at which electromagnetic waves propagate, remains constant, with a few exceptions, at 299,792,458 meters per second (186,282 miles per second), which is a very high rate of speed from our point of view.   This would indicate that the pressure of the "physical substance" within the primordial singularity remains constant, but that it is under extremely high pressure, or tension.

This evidence, combined with the fact that electromagnetic waves can have both a transverse polarization and a longitudinal polarization, can give us a strong indication of what conditions are really like within the primordial singularity which produces the universe that we see --- that it exists in a "fixed" state and under extremely high pressure or concentration.

In the next chapter, we will see how the conditions within the primordial singularity not only determine the speed at which electromagnetic waves propagate, but also our perceptions of mass (atoms).

CHAPTER 6

THE BASIS OF QUANTUM THEORY

By the end of the 1800's, great progress had been made within the realm of physical science. Newtonian physics was well established, as well as James Clerk Maxwell's theory of electricity and magnetism. The state of our knowledge of physics during this time period is referred to by physicists as **classical physics.** The new theories and ideas which emerged from the revolution in physical science at the turn of the century, became known as **modern physics**.

Toward the end of the 1800's, physicists were using new technology to devise more highly sophisticated experiments. They measured the speed of light with greater precision than ever before, and then they began to study the effects which relative motion had on the light (electromagnetic) waves such as the velocity of the Earth as it travels in its orbit around the sun. A number of serious contradictions began to surface, and it soon became clear that classical physics was running into trouble.

Another interesting development which took place in the early 1900's which created even further controversy among the leading physicists of that time period, was the development of quantum theory. In 1900, a German mathematics professor named Max Planck (1859 – 1947) worked out a formula which accurately described the intensities of electromagnetic waves at various frequencies which were emitted by a white-hot object. In order for the formula to work out correctly, a certain constant had to be included which eventually became known as Planck's Constant (see page 126).

One of the most surprising results of this historic development was that it suggested that electromagnetic wave energy was not emitted in an even and continuous manner, but actually in discreet amounts, or individual wavelets, called "quanta". Thus, the name Quantum Theory was derived.

Since this new theory dealt primarily with electromagnetic wave/particles, physicists recognized almost from the start that the new Quantum Theory was not entirely compatible with Einstein's Theory of Relativity, especially at the atomic level. As more advanced research was conducted and both theories were developed further, the inability of The Theory of Relativity to completely describe certain quantum processes regarding the "fixed" electric and magnetic field characteristics of the electromagnetic wave/particles became even more obvious.

In his Theory of Relativity, Dr. Albert Einstein used equations which described energy levels and mass in a continuous manner. Quantum Theory, however, used probabilities and matrices to describe specific energy states and decay processes regarding the electromagnetic wave/particles. As the mathematics of Quantum Theory was developed further, Planck's Constant was used more and more often in the mathematical descriptions of different energy states and other quantum processes. It soon became apparent that Planck's Constant was a very important component part of the universe in which we live.

One of the early successes of Quantum Theory was that it describes how energy is transferred at the atomic level. Thus, Quantum Theory deals primarily with electromagnetic wave/particles (atoms), and energy transfer among atoms by the means of electromagnetic waves. As a result, many of the contradictions which led to the demise of the Ether Theory, began to resurface while the new Quantum Theory was being developed.

When an electric or magnetic field is produced, it requires a certain amount of energy to cause it to build up. Then, the fields retain that energy as potential energy until the fields collapse, at which time, it is released as kinetic energy. The stronger that the electric and magnetic fields are in an electromagnetic wave, the more energy is stored in the amplitude of the wave. Therefore, the total amount of energy which is contained in, or can be transferred by, an electromagnetic wave is equal to the amplitude of the wave multiplied by its frequency.

One of the most important among the early discoveries which led to the development of Quantum Theory, indicated that all of the electromagnetic waves which were emitted from atoms have exactly the same amplitude which is represented by Planck's Constant, only their frequency varies. Thus, it was determined that the amplitude of the electromagnetic waves of which atoms consist, and also the electromagnetic waves which are produced by atoms, are all precisely the same, which is represented by Planck's Constant.

Whenever an atom releases energy in the form of electromagnetic waves, it creates a disturbance within the four-dimensional space-time energy field, which is represented by the Permittivity and Permeability values, which is similar to dropping a stone into a pool of water. Not just one, but a rather large number of individual wavelets are produced. The sum of all the wave energy which is transferred by all the individual wavelets, which were produced from the instant that the stone hit the water until the pool became quiet again, equals the total amount of energy which was released by the stone hitting the water.

When an electron changes energy states within an atom, for instance, this disturbance creates nearly a million individual electromagnetic waves in the four-dimensional space-time energy field which is represented by the Permittivity and Permeability values. With electromagnetic waves which are produced by atoms, or the component parts of atoms, the amplitude of all the waves are all the same (Planck's Constant). The frequency of the waves are what actually determines how much energy is transferred by the total energy release.

Earlier, we discussed how electromagnetic waves experience the maximum effects of relativity as a result of their propagation rate, and how this causes the electromagnetic waves to be contracted into a point particle. Since all of the electromagnetic waves which were produced by each event, were also created at the same point of origin (location) and were created at the same time, then they are all contained within the same point particle which is commonly referred to as a photon (see page 117).

What Planck's Constant actually represents is the small amount of energy which is contained by the amplitude of the electric and magnetic fields in each individual wavelet within the photon. Since the electromagnetic fields of which the amplitude of the wave consists are being held in a fixed state (called an "action"), the actual potential energy they represent is very, very small (see page 126). Also, the amplitude of the individual wavelets in the photon are exactly the same as the amplitude (which is represented by Planck's constant) of the electromagnetic wave/particle in the atom which created the disturbance within the four-dimensional space-time energy field.

Since all of the electromagnetic waves of which atoms consist have the same amplitude, and roughly the same frequency, this would make a rather strong argument that the electromagnetic waves of which atoms consist, were all produced in the same cosmic event. In this case, it would be the original electromagnetic shock wave which was in the primordial singularity when it came into our existence within the "empty black void". In all probability, all of the electromagnetic waves which originated within the primordial singularity, are actually clones of this original shock wave. That is why their amplitudes are all exactly the same.

As the primordial singularity expanded to its next step, it disappeared from our existence within the "empty black void" and went on to the next existence where it represents a different size and pressure, and so forth, until it expanded to the point where the "physical substance" was dissipated to the extent that it could no longer form atoms. From then on, the existences which came after that, were empty and unable to produce a universe like ours. The size and pressure differences which represent the "quantum jumps" from one existence to the next have actually been calculated (see page 140).

The four-dimensional space-time energy field which was established by the overall expansion process of the "physical substance" within the primordial singularity is what is actually producing the effects of relativity in all of these separate universes. Furthermore, the size and pressure of the primordial singularity which represents our existence within the void, remains "stuck" at that particular size (1.65 X 10^{-13} cm) in diameter and at a pressure which is represented by the Permittivity and Permeability values, which, in turn, causes electromagnetic waves (light) to propagate at a rate of 186,282 miles per second, or 299,792,458 meters per second (as we perceive it).

In universes which came into existence before ours, the inhabitants would perceive the speed of light to be greater, electromagnetic wave/particles (atoms) would appear smaller, and their whole universe would "appear" larger than ours. In universes which came into existence after ours, the inhabitants would perceive the speed of light to be slower, their atoms would appear to be larger, and their universe would appear smaller.

In the next separate existence within the void, the size and pressure of the primordial singularity are different, so the speed of light there would be different, and the size of the electromagnetic wave/particles would also be different within that universe. These separate universes remain separate from each other simply because energy cannot travel from one existence to the other. We are not aware of their existence because the electromagnetic (light) waves are not compatible with the atoms in our universe. In other words, the "fixed" effects of electricity and magnetism are not quite the same in any of these separate universes.

The means by which electromagnetic wave energy is transferred from one place to another within our universe is very important to us and our universe. Without it there would be no light, heat, energy, gravity, or mass. Without energy, our universe would be an empty, dark, void. Since energy itself, and also the transfer of energy, is such a very important part of our universe, then it must also be a very important part of the underlying physical process which is producing the universe that we see.

There are two major categories in which all forms of electromagnetic wave energy in our universe are included --- kinetic energy, potential energy. Kinetic energy is the energy of motion, and thus, we have the connection to the effects of relativity, velocity-related space-time distortion. Potential energy represents the electric and magnetic field energy which was established when they were built up, like compressing a couple of springs, and then holding them.

The potential energy in an electromagnetic wave is represented by the amplitude of the wave which is being held in a "fixed" state by the nature of the laws of physics in the "empty black void". In addition to this, most of the electromagnetic wave energy of which atoms consist (mass, which then produces gravity as a result) also falls into this category because these electromagnetic waves are "at resonance" and are being held as potential energy due to the nature of the laws of physics in the "empty black void".

If you could take an object with even a small amount of inertial mass and detonate it in a nuclear explosion, you would find that it consists of an enormous amount of electromagnetic wave energy which is released in the form of electromagnetic waves at all different frequencies. Even though a very large amount of kinetic energy is released in a nuclear explosion, only a relatively small amount of the total potential energy (mass) which is involved in the explosion would be actually converted into kinetic energy.

The enormous amount of potential electromagnetic wave energy which is represented by mass is described by Dr. Albert Einstein's famous equation (E = m c²), which means that the total amount of potential energy (E) contained within a certain amount of mass (m) is equal to that measured amount of mass multiplied times the speed of light squared. Consequently, electromagnetic wave energy is related to space-time distortion through its inertial mass, wavelength (space), and frequency (time interval).

A neutron, which represents the initial electromagnetic shock-wave in the primordial singularity, has a "Sphere of Influence" or wave function (Ψ) which **is** the size of the outer reaches of our universe. In other words, the "Sphere of Influence" of all the neutrons **are** the universe.

Thus the wave function (Ψ) simultaneously represents both the amplitude of the neutron electromagnetic wave at any point within our universe, and also the "probability amplitude" of finding the wave/particle existence of the neutron at any point within, or on the outer surface of, our universe.

Furthermore, every electromagnetic wave/particle which was created during the time of the initial shock-wave in the primordial singularity, has a "Sphere of Influence" (wave function) which is a small part of our whole universe and contributes to the collective strength of the Permittivity and Permeability values, right on down to the "Big Bang" background blackbody radiation which persists to the present-day.

So, according to this theory, Planck's constant " h " represents the amplitude of the initial electromagnetic shock-wave in the primordial singularity at the instant before it started expanding, and (Ψ) represents the amplitude of that same electromagnetic shock-wave as it has expanded and spread out to the size it is today (the size of our universe).

Now, let us consider a photon or "quantum of energy" which has been emitted from a wave/particle that has the primordial fireball as its point of origin. This photon will "inherit" the "Sphere of Influence" (wave function) of its parent wave/particle, in addition to its own, newly created, "Sphere of Influence".

This situation is what is causing all the faster-than-light responses, which have been observed in quantum experiments. Electromagnetic wave/particles "maintain" their existence back at their point of origin because of their electromagnetic characteristics, and partly due to the effects of relativity. So if you do something to one of these wave/particles, such as observe it, or if it is "captured" by another electromagnetic wave/particle, its existence is affected all the way back to its point of origin and all across its "Sphere of Influence" (wave function collapse).

This is why these high-frequency electromagnetic wave/particles have such a strange existence, so that the more you know about their location (space coordinates), the less you can know about their momentum (time coordinates), and that you can only tell where they are, what their momentum is, or what energy state they are in, just while you are actually observing them (Uncertainty Principle).

In 2004 a movie was released with the title "What The Bleep Do We Know" which highlights the strange existence of atomic wave/particles (which we are also made of) within our spacetime continuum. In 2004 a book of quotations from this movie was published by Captured Light Distribution, titled "The Little Book Of Bleeps", ISBN 0-9761076-0-4. The information in this movie and book demonstrates the "strangeness" of these atomic wave/particles. The physical constants which represent the electromagnetic characteristics of these wave/particles, also fit into a large mathematical framework which represents the underlying physical processes which produce the universe that we see (see pages 123 thru 145).

The fundamental interactions between electromagnetic waves/particles which enable atoms and molecules to exist are a fascinating aspect of quantum physics (Quantum Mechanics). One of the most important of these basic principles deals with the discoveries regarding how electrons form a "cloud of charge" around the nucleus of an atom. As we have seen previously, electrons which represent the negative half-wave of the initial shock-wave in the primordial singularity, have a frequency of around 10^{21} Hz.

Atoms and molecules function according to the principle of electromagnetic wave resonance, such that the diameters of each of the various electron "shells" in an atom, are determined by wave harmonics, or circular standing waves. Only electrons with the proper number of wavelengths, polarization of electric charge, polarization of magnetic fields, transverse or longitudinal wave polarization, orbital angular momentum, magnetic angular momentum, and spin, are able to share the various electron shells, and so no two

84

electrons in the same quantum state are allowed to occupy the same electron shell within an atom (Pauli Exclusion Principle).

The spin value is the quantum value of an electromagnetic wave/particle which represents the amount of energy consumed by the angular momentum of the physical substance as it "changed" from its lower pressure (-) state, to its higher pressure (+) state. This intrinsic angular momentum energy value is also held in a "fixed state" within the realm of "basic reality" and contributes to the physical, or "quantum", characteristics of an electromagnetic wave/particle within an atom.

The nuclei must also remain within specific wave orientations and energy levels in order to maintain stability. There are many more additional details regarding atoms and molecules to be found in Quantum Mechanics, however, it should be obvious to the reader by now that atoms and molecules, which are fundamental components of our universe as a whole, are comprised of electromagnetic waves at resonant frequencies, and that mass consists entirely of electromagnetic wave energy.

Because of these interrelationships, it becomes rather obvious that all of these various "effects" are being produced by a single underlying physical process (see page 118). As you can see, our perceptions of these different forms of energy are all being produced by the same underlying physical process (the primordial singularity) which is producing our perceptions of electric and magnetic fields in addition to our four-dimensional space-time energy field which is represented by the Permittivity and Permeability values that are particular to our individual universe (existence) within the "empty black void.

Thus, we enter the era of modern physics in which the underlying physical processes that produce the world around us (the universe that we see) are being revealed. All of the facts about electricity, magnetism, relativity, and quantum theory begin to fit together in a very simple and orderly fashion. Instead of continuously raising more questions, the results of the latest quantum experiments will now provide us with more and more evidence which can now support a Unified Theory.

It is very important for us to continue working to develop a Unified Field Theory. A Unified Field Theory will not only assist us in the development of new technologies here at home, but most importantly, it will enable us to further our space exploration projects, as we shall see in the next chapter.

CHAPTER 7

OUR UNIVERSE

According to our observations, our universe is so large that we can barely comprehend it --- approximately 30 billion light years across (from one side to the other). One light year is equal to the distance light travels in 1 year, which is equal to about 946,000,000,000,000,000 meters. Our galaxy is roughly 100,000 light years across, and contains several billion stars. Our nearest galaxy, the Andromeda galaxy, is 2,300,000 light years away, and also contains a very large number of stars. In addition to this, astronomers estimate that the galaxies may number in the billions, with new ones being discovered all the time as more and more powerful telescopes are being built.

Our universe is so large, that it provides us with almost unlimited opportunities for exploration and development. All we have to do in order to take advantage of this great opportunity, is to develop a satisfactory means of space travel so we can cross these vast distances within a reasonable amount of time.

Already, we have landed robotic explorers on several celestial bodies within our solar system, and we have made plans to explore several more. We are even thinking about sending astronauts to Mars, our nearest planet. Up until now, we have maintained an impressive record of achievement by using our present technology. However, with our current level of technology, it will still take around 3.5 years for a conventional spacecraft to reach Mars.

This is a wonderful goal for us to try to achieve, and with determination and perseverance, we can attain it. Our current means of space travel will probably continue serving us well into the 21st century, as long as we travel to the inner planets and moons within our solar system. But how are we going to travel beyond our solar system? This is going to prove to be very difficult with the type of space craft we have in use today. With our current type of space craft, it would be extremely difficult for us to reach the outer planets in our solar system which include: Jupiter, Saturn, Neptune, and Uranus.

Even if we were able to travel at the speed of light, it would still take us more than ten years to reach our nearest star, Alpha Centauri, let alone, our nearest galaxy. Let's face it, with the type of space craft we have in use today, our universe is virtually closed to our exploration. So in order for us to pursue the level of technological advancement which we have maintained throughout our history, we will be required to design and build a much better type of space craft.

Considering the great distances which we will need to travel, our new type of space craft design will have to be radically different from that which we are using now. Many times in the past, scientific achievement seemed to be grinding to a halt, with no hope of improving the situation. Then, someone found an entirely different approach to meet the challenge, and then a whole new era of scientific achievement usually followed.

In order to further our technological advancement, we are going to need to develop a comprehensive, Unified Field Theory which will give us a better understanding of how our universe really functions. Then we will be able to use this information to design and build a better spacecraft which will have capabilities many times greater than the ones in use today.

With the application of the principles in this book regarding electricity and magnetism, the possibility exists that the four-dimensional space-time energy field which causes us to perceive these great distances, can be distorted sufficiently in order to permit us to travel to distant places within our universe within a short period of time. And by designing a space craft which can produce space-time distortion by using electric and magnetic fields, we will be able to open up our entire universe for further exploration.

Surprisingly, some of this new technology may already exist, but it has been deemed classified and lies hidden away in a vault somewhere. One such experiment which may have actually achieved quantum teleportation/transportation has become known as the "Philadelphia Experiment" because it was allegedly conducted at the Philadelphia Navy Yard in 1943. It was suspected of involving the U.S. Navy destroyer DE – 173 (the USS Eldridge) see the book shown on page 119.

If the U.S. Navy records can be located and declassified regarding this alleged incident, this may turn out to be the first recorded event where quantum teleportation/transportation was actually achieved with a large craft (the ship). If this turns out to be the case, it will eventually become a major, historically significant event.

This project is suspected of being another invention of Nicola Tesla (1856 - 1943) who also invented our AC three-phase power distribution system, wireless transmission of power, the first cyclotron which was used for early nuclear research, the tesla coil, and fluorescent lighting to name a few. If the records can be declassified and we can prove that this experiment actually took place, Nikola Tesla should receive the credit for a historical event as significant as the Wright brothers' development of the first airplane or the historic first moon landing.

In the picture on page 120, Nikola Tesla is sitting beside one of his helical tesla coils which were suspected of being placed in a large cylindrical metal container, filled with transformer oil, and then bolted to the hull of the

DE – 173. This would enable the experimenters to charge the hull of the ship to many millions of volts. In addition to this, it is also suspected that a large electromagnet ran through the length of the ship. A special mast was suspected of being installed and outfitted with antennas which transmitted electromagnetic waves (radio waves) that were reflected from the sides of large cargo ships. These cargo ships were then positioned on a parallel heading approximately 100 meters away on each side and the electromagnetic waves were then brought to resonance between the cargo ships and the DE - 173.

Considering the amount of power which was suspected of being used, there is a fairly good chance that what the eyewitnesses claimed they saw, actually occurred. The way that this experiment was suspected of being set up, it would have been very dangerous to the experimenters and highly inefficient, but it still may have achieved the desired result. Modern-day experiments of this type utilize an enclosed resonant cavity design which is far more efficient. Some very interesting results can be obtained with an ordinary signal generator and a spectrum analyzer by using these devices.

Quantum teleportation has been demonstrated quite reliably on the atomic scale. Considering the hazards which may be involved, it is highly recommended that remote controls and robotics should be used exclusively during the early stages of this research. It is strongly suspected that Nicola Tesla objected vehemently to the use of live subjects during early trials of the Philadelphia Experiment, but his warnings probably went unheeded because it is doubtful that anyone who was involved in the project had any idea about what was really going on.

As it was with the early development of airplanes, there are going to be hazards. However, with the widespread use of remote controls, robotics, and common sense safety precautions, it is quite possible that this technology could be utilized to open up our entire universe for our exploration.

GLOSSARY

α ----- **fine structure constant** Introduced in 1916 by Arnold Sommerfeld (1868 – 1951) it represents the strength of the electromagnetic interaction between elementary charged particles.

abstract Considered apart from matter; imaginary. The essence of a larger object or whole.

astronomical Very large, as the numbers or quantities used in astronomy.

asymmetrical Not symmetrical, or, un-symmetrical. (see symmetrical)

axis A line around which a turning body rotates. A fixed line, as in a graph, along which distances are measured or to which positions are referred.

"basic reality" The "empty black void" in which our universe exists. There is no four-dimensional space-time energy field within this realm (except that which is contained within our universe), so the concepts of space and time, as we understand them, are meaningless beyond the outer reaches of our universe. For purposes of clarification, this realm was named "basic reality" because it represents what is actually real within our universe. Our perceptions of space and time are a grand-scale optical illusion which is being produced by our four-dimensional space-time energy field.

binary star system   A pair of stars which orbit around a common center of gravity.   Astronomers believe that most stars exist in this manner.

black hole   A theoretical entity formed by the explosion (supernova) and resulting contraction of a large star, greater than 2 ½ times the mass of the sun. A region of space where the effects of the four-dimensional space-time energy field are almost totally distorted (gravitational field) by the tremendous amount of mass within the singularity which lies at the center of the black hole.   According to this theory, total space-time distortion could be achieved by a black hole only if the total mass of the entire universe was once again contained within its singularity.

blackbody   An ideal physical body which would absorb all incident radiation falling on it including all the radiation it emits.

bulk modulus   The amount of force per unit area required to achieve a given amount of deformation.

chronometer   Precision clock for accurate measurements of time over long durations.

cosmology   (as it applies to astronomy and astrophysics) is the study of the structure, dynamics, and the development of our universe.   A cosmology theory attempts to explain how our universe was formed, what has happened to it in the past, and what might happen to it in the future.

Doppler effect The change, or shift, in the frequency of sound or light (electromagnetic) waves .

essence   That which makes something what it appears to be.

format   The general form or arrangement of anything.

four-dimensional space-time   In this theory, it is referred to as the abstract energy field which produces our perceptions (what we see) regarding the universe in which we live.   It is represented by two electrical terms known as the Permittivity of free space and the Permeability of free space (see page 30).

galaxy   A large cluster of stars, gas, dust, and, in many cases, a black hole.

homogeneous   Of the same kind, order, or nature. All the same.

inertial reference frame   A reference frame in which an object experiences no net force due to other bodies. A spacecraft travelling in a straight line at a constant velocity, not rotating, and far from any gravitational fields, would be an example.

inverse   A reversed or opposite orientation.   As one value becomes larger, the other becomes proportionately smaller and vice versa.

isotropic   Evenly spread.   Evenly distributed throughout.

kinetic   Of, or pertaining to motion.   Kinetic energy; energy of motion; energy in motion.

lumeniferous ether   Early physicists believed that this was the physical substance through which electromagnetic waves propagate (travel).

relative motion of the source of the waves either toward or away from the observer.

dynamic   Active;   Having to do with physical energy; Constantly changing;

effects   (of four-dimensional space-time) Impressions that we have about our universe and what we see around us.   Relativity is an accurate description of the effects of four-dimensional space-time and quantum theory is an accurate description of the interaction between the effects of four-dimensional space-time and the realm of "basic reality".

elasticity   Spontaneously returning to former shape after compression, extension, or other distortion.   The property or quality of being elastic.

electromagnetic wave polarization   The orderly alignment of the electric and magnetic fields of which an electromagnetic wave/particle consists.   Once the electric and magnetic fields are aligned in a particular orientation, they will be held in that state by the nature of the void "basic reality" in which our universe exists.

entropy   The principle , most commonly applied in Thermodynamics, which states that the complexity or disorder in a system increases as time proceeds.

EPR paradox   Written in 1935 by Albert Einstein, Boris Podolsky, and Nathan Rosen, it represented a hypothetical challenge to Quantum Mechanics that electromagnetic wave/particles could communicate with each other instantly across arbitrarily large distances of space – faster than the speed of light.

theory, there actually <u>is</u> a physical medium which produces our perceptions of electromagnetic waves, but it has a "real" existence only within the primordial singularity within our "Dimension" (universe) (see page 136). The actual physical characteristics of this substance are represented by the Permittivity and Permeability of free space.

molecule   Two or more atoms joined together.

negative space-time distortion   The space and time relationship which would be experienced by two objects which have a relative motion <u>away</u> from each other, and an object in a gravitational field.   The Permittivity and Permeability values would be reduced.

paradox   A statement which seems to contradict itself, but which may be true.

Permeability of free space   A value of proportionality that exists between magnetic flux density and magnetic field strength.   (μ_0) the Permeability of free space: 1.256,637,061,44E-7 Henry's per meter^{-1} (The $^{-1}$ exponent means that there is an inverse relationship between ε_0 and μ_0)

Permittivity of free space   A value of proportionality that exists between electric flux density and electric field intensity.   (ε_0) the Permittivity of free space: (8.854,187E-12 Farads per meter)

perpendicular   At a right angle to.

point of "basic reality"   The point at which our concepts of space and time become meaningless.   A black hole singularity, a neutron, or something travelling at the speed of light all represent a reference frame in which the effects of the four-dimensional space-time energy field are distorted down to very nearly zero, or 1.65×10^{-13} cm.

positive space-time distortion   The space and time relationship which would be experienced by two objects which have a relative motion <u>toward</u> each other.   The Permittivity and Permeability values would be increased.

primordial singularity   The small (1.65×10^{-13} cm), highly compressed singularity which is held in a "fixed state" within the realm of "basic reality".   The singularity has actually expanded in other dimensions, or existences, to create the four-dimensional space-time energy field which produces our universe that we see today.

propagate   Travel through, as in wave motion.

proportionately   Symmetrical in relative magnitude or arrangement.   Equality of ratios.

quanta   Discrete amounts of electromagnetic wave energy or energy states associated with atomic interactions.   Electromagnetic wavelets carrying energy equal to their amplitude multiplied by their frequency.

quantum entanglement   A phenomenon in which two or more separate electromagnetic wave/particles are interactive and instantaneous across arbitrarily large distances of space.   If a measurement is made on one of the wave/particles, the other(s) will "know" instantly what measurement has been made and also the outcome of that measurement.

quantum event   Any electromagnetic disturbance which involves waves in the four-dimensional space-time energy field (represented by the Permittivity and Permeability of free space).

solar system   The Sun, including the planets, with orbiting moons,

spectrum   An orderly display of electromagnetic radiation which has been diffracted and arranged in the order of the respective wavelengths, or frequencies, of which the radiation consists.

sphere   A geometrical form in which all points on the surface are equidistant from the center point.   In this theory, the center point of a sphere can also represent all points on the surface, and all points within the sphere simultaneously (quantum mechanics).

"sphere of influence"   The space component, or wave function (as we see it) of an electromagnetic wave as it expands outward in all directions at the speed of light.

spring constant   The measure of "stiffness" or "elasticity" within an object, a particular substance, or medium which is capable of producing simple harmonic motion.

stellar   of the stars, or of a star

supernova   A star which contains at least 2½ times the mass of our Sun, and which burns its hydrogen and helium much more rapidly than a less-massive star due to its increased gravitational pressure.   Its burning process continues to produce heavier and heavier elements until it starts producing significant quantities of uranium. It first collapses, and then it undergoes a huge explosion in which the core can be ripped apart to form planets and moons, or it can condense even further to form a black hole.

symmetrical   A condition in which one part is proportionate to another. Composed of two parts which correspond harmoniously to each other.

viscosity   A measure of a fluid's resistance to flow or movement which may also be referred to as "thickness".

wave/particle   An electromagnetic wave in four-dimensional space-time which propagates through space at the speed of light. It can also appear as a particle due to its existence within the primordial singularity at the point of "basic reality".

BIBLIOGRAPHY

These are some of the books you may consider reading which provide informational support and to help you envision the bigger picture so you can see how all this information fits together.   By using this theory as an outline, you can have many wonderful hours of interesting reading while piecing together the observations and experimental results which are presented clearly and concisely in each of these great works.

Alexandrov, Yu. A..   **Fundamental Properties of the Neutron**.   Oxford: Oxford University Press,   1992.

Boslough, John.   **Masters of Time.**   New York: Addison-Wesley, 1992.

Boslough, John,   **"Searching for the Secrets of Gravity"**. National Geographic, 175:5, May 1989.

Calder, Nigel.   **Einstein's Universe.**   New York: Greenwich House, 1979.

Cohen-Tannoudji, Gilles.   **Universal Constants in Physics**.   New York: McGraw-Hill, Inc.,   1991.

Coveney, Peter, and Roger Highfield.   **The Arrow of Time.** New York: Fawcett Columbine, 1990.

Davies, Paul, and Gribbin, John. **The Matter Myth.** New York: Simon & Shuster, 1992.

Dirac, P. A. M. **The Principles of Quantum Mechanics.** New York: Oxford University Press, 1958.

Einstein, Albert. **RELATIVITY** **The Special and the General Theory.** New York: Crown Publishers, Inc., 1961.

Einstein, Albert; Lorentz, H. A.; Minkowski, H.; and Weyl, H. **The Principle of Relativity.** New York: Dover Publications, Inc., 1952.

Ellis, George F. R., and Williams, Ruth M. **Flat and Curved Space-Times.** New York: Oxford University Press, 1988.

Gazdag, László. **Beyond the Theory of the Relativity.** Budapest: Ernő Werőczei, 1998.

Giancoli, Douglas C. **General Physics.** Englewood Cliffs, New Jersey: Prentice-Hall, 1984.

Gribbin, John. **In Search of Schrödinger's Cat.** New York: Bantam, 1984.

Hawking, Stephen W. **A Brief History of Time.** New York: Bantam, 1988.

Herbert, Nick. **Quantum Reality.** New York: Anchor Books, 1985.

Horgan, John. **"Quantum Philosophy".** Scientific American, July, 1990.

Kaufman, Nicholas J. **Black Holes and Warped Space-time.** W. H. Freeman and Company, 1979.

Macrae, Norman.   **John von Neumann**.   New York: Pantheon Books,   1992.

Marion, Jerry B.   **Our Physical Universe.**   New York: John Wiley & Sons, Inc., 1971.

Martin, Thomas C..   **The Inventions, Researches, and Writings of Nikola Tesla**.   New York: Barnes & Noble Books,   1992.

Matlin, Marlee; Hendrix, Elaine; Bailey, Robert, Jr.. and numerous other authors.   **The Little Book of BLEEPS.** USA: Captured Light Distribution,   2004.

Maxwell, James Clerk.   **A Treatise on Electricity and Magnetism**.   Oxford: Clarendon Press,   1891.

Moore, William L. in consultation with Berlitz, Charles. **The Philadelphia Experiment:.**   New York: Grosset & Dunlap Publishers,   1979.

Murdin, Paul.   **End in Fire**.   New York: Cambridge University Press,   1990.

Pais, Abraham.   **Niels Bohr's Times, in Physics, Philosophy, and Polity**.   Oxford: Oxford University Press, 1991.

Peat, F. David.   **Einstein's Moon.**   Chicago: Contemporary Books, 1990.

Pierls, Rudolph. **More Surprises in Theoretical Physics.** Princeton, New Jersey: Princeton University Press, 1991.

Schrödinger, Erwin. **Space-Time Structure.** New York: Cambridge University Press, 1950.

APPENDIX

Twelve Star Publishing : Sourceworks

Ether and the Theory of Relativity

(An address delivered on May 5th, 1920, at the University of Leyden, The Netherlands; by Dr. Albert Einstein)

How does it come about that alongside of the idea of ponderable matter, which is derived by abstraction from everyday life, the physicists set the idea of the existence of another kind of matter, the ether? The explanation is probably to be sought in those phenomena which have given rise to the theory of action-at-a-distance, and in the properties of light which have led to the undulatory theory. Let us devote a little while to the consideration of these two subjects.

Outside of physics we know nothing of action-at-a-distance. When we try to connect cause and effect in the experiences which natural objects afford us, it seems at first as if there were no other mutual actions than those of immediate contact, e.g. the communication of motion by impact, push and pull, heating or inducing combustion by means of a flame, etc. It is true that even in everyday experience weight, which is in a sense action-at-a-distance, plays a very important part. But since in daily experience the weight of bodies meets us as something constant, something not linked to any cause which is variable in time or place, we do not in everyday life speculate as to the cause of gravity, and therefore do not become conscious of its character as action-at-a-distance. It was Newton's theory of gravitation that first assigned a cause for gravity by interpreting it as action-at-a-distance, proceeding from masses. Newton's theory is probably the greatest stride ever made in the effort towards the casual nexus of natural phenomena. And yet this theory evoked a lively sense of discomfort among Newton's contemporaries, because it seemed to be in conflict with the principle springing from the rest of experience, that there can be reciprocal action only through contact, and not through immediate action-at-a-distance.

It is only with reluctance that man's desire for knowledge endures a dualism of this kind. How was unity to be presented in his comprehension of the forces of nature? Either by trying to look upon contact forces as being themselves distant forces which admittedly are observable only at a very small distance—and this was the road which Newton's followers, who were entirely under the spell of his doctrine, mostly preferred to take; or by assuming that the Newtonian action-at-a-distance is only *apparently* immediate action-at-a-distance, but in truth is conveyed by a medium permeating space, whether by movements or by elastic deformation of this medium. Thus the endeavor toward a unified view of the nature of forces leads us to the hypothesis of an ether. This hypothesis, to be sure, did not at first bring with it any advance in the theory of gravitation or physics generally, so that it became customary to treat Newton's law of force as an axiom not further reducible.

But the ether hypothesis was bound always to play some part in physical science, even if at first only a latent part.

When in the first half of the nineteenth century the far-reaching similarity was revealed which subsists between the properties of light and those of elastic waves in ponderable bodies, the ether hypothesis found fresh support. It appeared beyond question that light must be interpreted as a vibratory process in an elastic, inert medium filling up universal space. It also seemed to be a necessary consequence of the fact that light is capable of polarization, that this medium, the ether, must be of the nature of a solid body, because transverse waves are not possible in a fluid, but only in a solid. Thus the physicists were bound to arrive at the theory of the "quasi-rigid" lumeniferous ether, the parts of which can carry out no movements relatively to one another except the small movements of deformation which correspond to light-waves.

This theory – also called the theory of the stationary lumeniferous ether – moreover found a strong support in an experiment which is also of fundamental importance in the special theory of relativity, the experiment of Fizeau, from which one was obliged to infer that the lumeniferous ether does not take part in the movements of bodies. The phenomenon of aberration also favoured the theory of the quasi-rigid ether.

The development of the theory of electricity along the path opened up by Maxwell and Lorentz gave the development of our ideas concerning the ether quite a peculiar and unexpected turn. For Maxwell himself, the ether indeed still had properties which were purely mechanical although of a much more complicated kind than the mechanical properties of tangible solid bodies. But neither Maxwell nor his followers succeeded in elaborating a mechanical model for the ether which might furnish a satisfactory mechanical interpretation of Maxwell's laws of the electromagnetic field. The laws were clear and simple, the mechanical interpretations clumsy and contradictory. Almost imperceptibly the theoretical physicists adapted themselves to a situation which, from the standpoint of their mechanical programme, was very depressing. They were particularly influenced by the electro-dynamical investigations of Heinrich Hertz. For whereas they previously had required of a conclusive theory that it should content itself with the fundamental concepts which belong exclusively to mechanics (e.g. densities, velocities, deformations, stresses) they gradually accustomed themselves to admitting electric and magnetic force as fundamental concepts side by side with those of mechanics, without requiring a mechanical interpretation for them. Thus the purely mechanical view of nature was gradually abandoned.

But this change led to a fundamental dualism which in the long-run was unsupportable. A way of escape was now sought in the reverse direction, by reducing the principles of mechanics to those of electricity, and this is especially as confidence in the strict validity of the equation of Newton's mechanics was shaken by the experiments with *beta*-rays and rapid cathode rays.

This dualism still confronts us in unextenuated form in the theory of Hertz, where matter appears not only as the bearer of velocities, kinetic energy, and mechanical pressures, but also as the bearer of electromagnetic fields. Since such fields also occur *in vacuo* – i.e. in free ether – the ether also appears as bearer of electromagnetic fields.

The ether appears indistinguishable in its functions from ordinary matter. Within matter, it takes part in the motion of matter, and in empty space it has everywhere a velocity; so that the ether has a definitely assigned velocity throughout the whole of space. There is no fundamental difference between Hertz's ether and ponderable matter (which in part subsists in the ether).

The Hertz theory suffered not only from the defect of ascribing to matter and ether, on one hand mechanical states, and on the other hand electrical states, which do not stand in any conceivable relation to each other, it was also at variance with the result of Fizeau's important experiment on the velocity of the propagation of light in moving fluids, and with other established experimental results.

Such was the state of things when H. A. Lorentz entered upon the scene. He brought theory into harmony with experience by means of a wonderful simplification of theoretical principles. He achieved this, the most important advance in the theory of electricity since Maxwell, by taking from the ether its mechanical, and from matter its electromagnetic qualities. As in empty space, so too in the interior of material bodies, the ether, and not matter viewed atomistically, was exclusively the seat of electromagnetic fields. According to Lorentz the elementary particles of matter alone are capable of carrying out movements; their electromagnetic activity is entirely confined to the carrying of electrical charges. Thus Lorentz succeeded in reducing all electromagnetic happenings to Maxwell's equations for free space.

As to the mechanical nature of the Lorentzian ether, it may be said of it, in a somewhat playful spirit, that immobility is the only mechanical property of which it has not been deprived by H. A. Lorentz. It may be added that the whole change in the conception of the ether which the special theory of relativity brought about, consisted in taking away from the ether its last mechanical quality, namely, its immobility. How this is to be understood will forthwith be expounded.

The space-time theory and the kinematics of the special theory of relativity were modelled on the Maxwell-Lorentz theory of the electromagnetic field. This theory therefore satisfies the conditions of the special theory of relativity, but when viewed from the latter it acquires a novel aspect. For if K be a system of co-ordinates relatively to which the Lorentzian ether is at rest, the Maxwell-Lorentz equations are valid permanently with reference to K. But by the special theory of relativity, the same equations, without any change of meaning, also hold in relation to any new system of co-ordinates K' which is moving in uniform translation relatively to K. Now comes the anxious question: -- Why must I in the theory distinguish the K system above all K' systems, which are physically equivalent to it in all respects, by assuming that the ether is at rest relatively to the K system? For the theoretician such an asymmetry in the theoretical structure, with no corresponding asymmetry in the system of experience, is intolerable. If we assume the ether to be at rest relatively to K, but in motion relatively to K', the physical equivalence of *K* and *K'* seems to me from the logical standpoint, not indeed downright incorrect, but nevertheless inacceptable.

The next position which it was possible to take up in the face of this state of things appeared to be the following. The ether does not exist at all. The electromagnetic fields are not states of a medium, and are not bound down to any bearer, but they are independent realities which are not reducible to anything else, exactly like the atoms of ponderable matter. This conception suggests itself the more readily as, according to Lorentz's theory, electromagnetic radiation, like ponderable matter, brings impulse and energy with it, and as, according to the special theory of relativity, both matter and radiation are but special forms of distributed energy, ponderable mass losing its isolation and appearing as a special form of energy.

More careful reflection teaches us, however, that the special theory of relativity does not compel us to deny the ether. We may assume the existence of an ether, only we must give up ascribing a definite state of motion to it, i.e. we must by abstraction take from it the last mechanical characteristic which Lorentz had still left it. We shall see later that this point of view, the conceivability of which I shall at once endeavor to make more intelligible by a somewhat halting comparison, is justified by the results of the general theory of relativity.

Think of waves on the surface of water. Here we can describe two entirely different things. Either we may observe how the undulatory surface forming the boundary between water and air alters in the course of time; or else – with the help of small floats, for instance – we can observe how the position of the separate particles of water alters during the course of time. If the existence of such floats for tracking the motion of the particles of a fluid were a fundamental impossibility in physics – if, in fact, nothing else whatever were observable than the shape of the space occupied by the water as it varies in time, we should have no ground for the assumption that water consists of moveable particles. But all the same we could characterize it as a medium.

We have something like this in the electromagnetic field. For we may picture the field to ourselves as consisting of lines of force. If we wish to interpret these lines of force to ourselves as something material in the ordinary sense, we are tempted to interpret the dynamic processes as motions of these lines of force, such that each separate line of force is tracked through the course of time. It is well known, however, that this way of regarding the electromagnetic field leads to contradictions.

Generalizing we must say this: -- There may be supposed to be extended physical objects to which the idea of motion cannot be applied. They may not be thought of as consisting of particles which allow themselves to be separately tracked through time. In Minkowski's idiom this is expressed as follows: -- Not every extended conformation in the four dimensional world can be regarded as composed of world threads. The special theory of relativity forbids us to assume the ether to consist of particles observable through time, but the hypothesis of ether in itself is "not" in conflict with the special theory of relativity. Only we must be on our guard against ascribing a state of motion to the ether.

Certainly, from the standpoint of the special theory of relativity, the ether hypothesis appears at first to be an empty hypothesis. In the equations of the electromagnetic field there occur, in addition to the densities of the electric charge, *only* the intensities of the field. The career of electromagnetic processes *in vacuo* appears to be completely determined by (these) equations, uninfluenced by other physical quantities. The electromagnetic fields appear as ultimate, irreducible realities, and at first it seems superfluous to postulate a homogeneous, isotropic ether-medium, and to envisage electromagnetic fields as states of this medium.

But on the other hand there is a weighty argument to be adduced in favor of the ether hypothesis. To deny the ether is ultimately to assume that empty space has no physical qualities whatever. The fundamental facts of mechanics do not harmonize with this view. For the mechanical behavior of a corporeal system hovering freely in empty space depends not only on relative positions (distances) and relative velocities, but also on its state of rotation, which physically may be taken as a characteristic not appertaining to the system itself. In order to be able to look upon the rotation of the system, at least formally, as something real, Newton objectivises space. Since he classes his absolute space together with real things, for him rotation relative to an absolute space is also something real. Newton might no less well have called his absolute space "Ether"; what is essential is merely that besides observable objects, another thing, which is not perceptible, (must) be looked upon as real, to enable acceleration or rotation to be looked upon as something real.

It is true that Mach tried to avoid having to accept as real something which is not observable by endeavoring to substitute in mechanics a mean acceleration with reference to the totality of the masses in the universe in place of an acceleration with reference to absolute space. But inertial resistance opposed to relative acceleration of distant masses presupposes action-at-a-distance; and as the modern physicist does not believe that he may accept this action-at-a-distance, he comes back once more, if he follows Mach, to the ether, which has to serve as (the) medium for the effects of inertia. But this conception of the ether to which we are led by Mach's way of thinking differs essentially from the ether as conceived by Newton, by Fresnel, and by Lorentz. Mach's ether not only *conditions* the behavior of inert masses, but is *also conditioned* in its state by them.

Mach's idea finds its full development in the ether of the general theory of relativity. According to this theory the metrical qualities of the continuum of space-time differ in the environment of different points of space-time, and are partly conditioned by the matter existing outside of the territory under consideration. This space-time variability of the reciprocal relations of the standards of space and time, or, perhaps, the recognition of the fact that empty space in its physical relation is neither homogenous nor isotropic compelling us to describe its state by ten functions (the gravitational potentials g, *mu, nu*), has, I think, finally disposed of the view that space is physically empty. But therewith the conception of the ether has again acquired an intelligible content, although this content differs widely from that of the ether of the mechanical undulatory theory of light. The ether of the general theory of relativity is a medium which is itself devoid of all mechanical and kinematical qualities, but helps to determine mechanical (and electromagnetic) events.

What is fundamentally new in the ether of the general theory of relativity as opposed to the ether of Lorentz consists in this, that the state of the former is at every place determined by connections with matter and the state of the ether in neighboring places, which are amenable to law in the form of differential equations; whereas the state of the Lorentzian ether in the absence of electromagnetic fields is conditioned by nothing outside itself, and is everywhere the same. The ether of the general theory of relativity is transmuted conceptually into the ether of Lorentz if we substitute constants for the functions of space which describe the former, disregarding the causes which condition its state. Thus we may also say, I think, that the ether of the general theory of relativity is the outcome of the Lorentzian ether, through relativation.

As to the part which the new (conception of the ether) is to play in the physics of the future we are not yet clear. We know that it determines the metrical relations in the space-time continuum, e.g. the configurative possibilities of solid bodies as well as the gravitational fields, but we do not know whether it has an essential share in the structure of the electrical elementary particles constituting matter. Nor do we know whether it is only in the proximity of ponderable masses that its structure differs essentially from that of the Lorentzian ether; whether the geometry of spaces of cosmic extent is approximately Euclidean. But we can assert by reason of the gravitation that there must be a departure from Euclidean relations, with spaces of cosmic order of magnitude if there exists a positive mean density, no matter how small, of the matter in the universe. In this case the universe must of necessity be spatially unbounded and of finite magnitude, its (magnitude) being determined by the value of that mean density.

If we consider the gravitational field and the electromagnetic field from the standpoint of the ether hypothesis, we find a remarkable difference between the two. There can be no space nor any part of space without gravitational potentials; for these confer upon space its metrical qualities, without which it cannot be imagined at all. The existence of the gravitational field is inseparably bound up with the existence of space. On the other hand a part of space may be very well imagined without an electromagnetic field; thus in contrast with the gravitational field, the electromagnetic field seems to be only secondarily linked to the ether, the formal nature of the electromagnetic field being as yet in no way determined by that of gravitational ether. From the present state of theory it looks as if the electromagnetic field, as opposed to the gravitational field, rests upon an entirely new formal *motif*, as though nature might just as well have endowed the gravitational ether with fields of quite another type, for example, with fields of a scalar potential, instead of fields of the electromagnetic type.

Since according to our present conceptions the elementary particles of matter are also, in their essence, nothing else than condensations of the electromagnetic field. Our present view of the universe presents two realities which are completely separated from each other conceptually, although connected causally, namely, gravitational ether and electromagnetic field, or, as they might also be called(,) space and matter.

Of course it would be a great advance if we could succeed in comprehending the gravitational field and the electromagnetic field together as one unified conformation. Then for the first time the epoch of theoretical physics founded by Faraday and Maxwell would reach a satisfactory conclusion. The contrast between the ether and matter would fade away, and, through the general theory of relativity, the whole of physics would become a complete system of thought, like geometry, kinematics, and the theory of gravitation. An exceedingly ingenious attempt in this direction has been made by the mathematician H. Weyl, but I do not believe that his theory will hold its ground in relation to reality. Further, in contemplating the immediate future of theoretical physics, we ought not unconditionally to reject the possibility that the facts comprised in the quantum theory may set bounds to the field theory beyond which it cannot pass.

Recapitulating, we may say that according to the general theory of relativity, space is endowed with physical qualities, in this sense, therefore, there exists an ether.

According to the general theory of relativity space without ether is unthinkable. For in such space there would be no propagation of light, but also no possibility of existence for standards of space and time (measuring-rods and clocks), not therefore, any space-time intervals in the physical sense. But this ether may not be thought of as endowed with the quality characteristic of ponderable media, as consisting of parts which may be tracked through time. The idea of motion may not be applied to it.

PHOTON

(two-dimensional view)

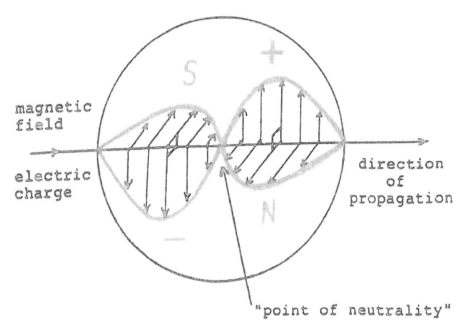

"point of neutrality"

A photon is an electro-magnetic wave/particle in four-dimensional space-time which has a dual existence... both wave and particle (as we see it). Its wave characteristics (wave function) travel outward in all directions at the speed of light (186,282 miles per second) like an expanding sphere. Its particle characteristics represent the electro-magnetic wave as it exists at the point of "basic reality" (primordial fireball singularity).

The point of "basic reality" (on the Time Chart) represents our universe as it exists in an otherwise empty and endless black void where time stands still (realm of "basic reality"). Our universe still exists there as a single point (primordial fireball singularity) as if the "big bang" has not occurred yet.

As we have seen earlier, the physical (electro-magnetic) properties of four-dimensional space-time exist only at the point of "basic reality", and the "pressure" of four-dimensional space-time inside the primordial fireball singularity was very, very high - so high, that electro-magnetic waves (photons) travel through it at a very high rate of speed (186,000 miles per second).

118

EXPANSION OF THE PHYSICAL SUBSTANCE

© 1995 by Charles R. Storey

Empty Black Void

LUMINIFEROUS ETHER (CLOUD) WHICH
PRODUCES OUR UNIVERSE THAT WE SEE
(WHERE IT IS NOW)

Empty Black Void

$S = 4\pi R^2$ THIS EQUATION DESCRIBES THE
SPHERICAL SHAPE OF THIS (CLOUD) OF
LUMINIFEROUS ETHER SUBSTANCE

$\dfrac{1}{R^2}$ LAW

Describes the physical expansion of the luminiferous ether (cloud) which produced the four-dimensional space-time power field.

Fitzgerald - Lorentz transformation equations:

$$T' = \frac{T}{\sqrt{1-\left(\frac{v}{c}\right)^2}}$$

$$L' = L\sqrt{1-\left(\frac{v}{c}\right)^2}$$

These equations taken together, actually describe distortions in the four-dimensional space-time power field which produces our perceptions of space, time, relative motion, acceleration and gravity.

The four dimensional space-time power field which produces our perceptions of time and space is caused by the tremendous pressure difference of the luminiferous ether between its starting point (primordial fireball singularity) and its expanded form (where it is now)

OUR DIMENSION (EXISTENCE)

$c = \sqrt{\dfrac{1}{\epsilon_0 \mu_0}}$

Starting point

The black dotted lines represent different (dimensions) existences of the luminiferous ether cloud. These separate existences are caused by the strange nature of the empty black void in which our universe exists. All of these separate existences exist as concentric spheres in all of their respective sizes and pressures simultaneously, as if they are "stuck" within the empty black void. They were drawn here in this manner to show that all of the different "existences" are separate from the others.

This equation represents the pressure/density of the luminiferous ether in the primordial fireball singularity which represents our dimension (existence)

The Philadelphia Experiment:

Project Invisibility

By William L Moore

In consultation with

Charles Berlitz

ISBN 0-448-15777-2

The speed of light $c = \dfrac{\alpha\hbar}{e^2} = \dfrac{1}{\sqrt{\varepsilon_0\mu_0}} = 299{,}792{,}458$ meters per second

1/the speed of light $\dfrac{1}{c} = \dfrac{e^2}{\alpha\hbar} = \sqrt{\varepsilon_0\mu_0} = 3.335{,}640{,}951{,}981{,}52\text{E-9}$ m/sec

Speed of light squared $c^2 = \dfrac{\alpha^2\hbar^2}{e^4} = \dfrac{1}{\varepsilon_0\ \mu_0} = 89{,}875{,}517{,}873{,}681{,}764$ m/sec

1/the speed of light squared $\dfrac{1}{c^2} = \dfrac{e^4}{\alpha^2\hbar^2} = \varepsilon_0\ \mu_0 = 1.112{,}650{,}056{,}053{,}62\text{E-17}$ m/sec

Coulomb constant $k_e = \dfrac{\alpha^2\hbar^2}{e^4} = \dfrac{1}{\varepsilon_0\ \mu_0} = 8.987{,}551{,}787{,}368{,}176{,}4\text{E9}$ N $\cdot$ m^{2}C^{-1}

$299{,}792{,}458 = \dfrac{1}{3.335{,}640{,}951{,}981{,}52} = \sqrt{89{,}875{,}517{,}873{,}681{,}764} = \dfrac{1}{\sqrt{1.112{,}650{,}056{,}053{,}62}}$

$3.335{,}640{,}951{,}981{,}52 = \dfrac{1}{299{,}792{,}458} = \sqrt{1.112{,}650{,}056{,}053{,}62} = \dfrac{1}{\sqrt{89{,}875{,}517{,}873{,}681{,}764}}$

$1.112{,}650{,}056{,}053{,}62 = \dfrac{1}{89{,}875{,}517{,}873{,}681{,}764} = 3.335{,}640{,}951{,}981{,}52^2 = \dfrac{1}{\sqrt{299{,}792{,}458}}$

$89{,}875{,}517{,}873{,}681{,}764 = \dfrac{1}{1.112{,}650{,}056{,}053{,}62} = 299{,}792{,}458^2 = \dfrac{1}{\sqrt{3.335{,}640{,}951{,}981{,}52}}$

*ESU - Electrostatic System of Units

ESU of Potential = 299,792,458/1.000,000 = 299.792,458 Volts
ESU of Resistance = 299,792,458^2/100,000 = 898,755,178,736.817,64 Ohms
ESU of Current = 1/299,792,458/10 = 3.335,640,951,981,52E-10 Amperes
ESU of Conductance = (1/299,792,458^2) X 100,000 = 1.112,650,056,053,62E-12 Siemens
ESU of Capacitance = 1/299,792,458^2 X 100,000 = 1.112,650,056,053,62E-12 Farads
ESU of Inductance = 299,792,458^2/100,000 = 898,755,178,736.817,64 Henrys
ESU of Quantity = 1/299,792,458/10 = 3.335,640,951,981,52E-10 Coulombs

---

Volt = 1/(299,792,458/1,000,000) = 3.335,640,951,981,52E-03 ESU of Potential
Ohm = 1/(299,792,458^2/100,000) = 1.112,650,056,053,62E-12 ESU of Resistance
Amperes = 1/(1/299,792,458)/10 = 2,997,924,580 ESU of Current
Siemens = 1(1/299,792,458^2 X 100,000 = 898,755,178,736.817,64 ESU of Conductance
Farads = 1/(1/299,792,458^2 X 100,000) = 898,755,178,736.817,64 ESU of Capacitance
Henrys = 1/(299,792,458^2/100,000) = 1.112,650,056,053,62E-12 ESU of Inductance
Coulombs = 1/(1/299,792,458)/10 = 2,997,924,580 ESU of Quantity

HIGH-PRECISION TABLE OF CONSTANTS

The six basic constants: c; e^2; α; $\hbar$; ε_0; and μ_0 appear to represent the basic internal structure and state of the physical substance within the primordial singularity immediately prior to its expansion, and therefore, may be considered as dimensionless during this stage of the process.

The instant that the singularity began to expand, the scalar properties of space, time, acceleration, and relative motion/heat came into existence which added increasing levels of complexity (thermal entropy) to the values which the higher-level constants represent -- with the six basic constants still representing the underlying structure from which the higher-level constants are patterned.

The Permittivity (ε_0) and Permeability (μ_0) of free space then represent these quantities into our reality (our perception) in their respective proportions and forms which include: meters per second; meters per second^2; Coulombs of electric charge; Farads of capacitance; Henry's of inductance; etc. In some cases, the exponents and units have been intentionally omitted since these values may be numerically representing several different quantities.

In many of the following examples, the calculated values of the physical constants may vary from the experimentally measured values by a few decimal places. Consequently, the primary objective here is to suggest the existence of an underlying physical framework which will ultimately provide us with the means of accurately calculating the precise values of all physical constants.

The values of the physical constants which are represented here are only to be used as examples. **They are not intended to replace the currently established values which have been determined by experiment.**

By setting the speed of light to exactly c = 299,792,458 meters per second, the precise values of ε_0 and μ_0 are now established as a metric standard by which all of the other physical constants may be measured. The precise values of each of the six basic constants may now be determined to a high degree of accuracy by definition which will enable us to calculate all of the physical constants to a high level of accuracy. Consequently, this will enable us to demonstrate how all the constants related to our spacetime continuum, the components of atoms, quantum physics, and electricity and magnetism are all interconnected through a comprehensive mathematical framework.

Physical orientations of the six basic constants:

$$c = \frac{\alpha \, \hbar}{e^2} \;\; ; \;\; e^2 = \frac{\alpha \, \hbar}{c} \;\; ; \;\; \alpha = \frac{c \, e^2}{\hbar} \;\; ; \;\; \hbar = \frac{c \, e^2}{\alpha} \;\; ; \;\; \varepsilon_0 = \frac{e^2}{\alpha \, \hbar \, c \, \mu_0} \;\; ; \;\; \mu_0 = \frac{e^2}{\alpha \, \hbar \, c \, \varepsilon_0}$$

Permittivity of free space:

$$\varepsilon_0 \;\; = \;\; \frac{e^4}{\alpha^2 \hbar^2 \mu_0} \;\; = \;\; \frac{e^2}{\alpha \, \hbar \, c \, \mu_0} \;\; = \;\; \frac{1}{c^2 \mu_0} \;\; = \;\; \frac{1}{4\pi c^2}$$

8.854,187,817,620,389,850,536,563,031,710,9e-12 Farads/meter

Permeability of free space:

$$\mu_0 \;\; = \;\; \textbf{12.566,370,614,359,172,953,850,573,533,118e-7 Henry's/meter}^{-1}$$

$$\mu_0 \;\; = \;\; \frac{e^4}{\alpha^2 \hbar^2 \varepsilon_0} \;\; = \;\; \frac{e^2}{\alpha \, \hbar \, c \, \varepsilon_0} \;\; = \;\; \frac{1}{c^2 \varepsilon_0} \;\; = \;\; \frac{2h}{\hbar} \quad \text{--- numerically equal to } 4\pi \text{ ---}$$

Impedence of free space:

$$Z_0 \;\; = \;\; c \, \mu_0$$

Conductance quantum: G_0 = 7.748,091,734,636,263,793,631,291,999,104,2

$$G_0 = \frac{2e^2}{h} = \frac{e^2}{\pi\hbar} = \frac{\alpha}{\pi c} = \alpha\sqrt{\varepsilon_0\mu_0} = \alpha \, c \, \varepsilon_0\mu_0 = \sqrt{8c}\,\sqrt{2c}\,\alpha\varepsilon_0 = \frac{\sqrt{\pi\varepsilon_0}\,\sqrt{8c}\,\sqrt{2c}\,e^2}{m_u\lambda_p \, c}$$

$$= \frac{\sqrt{\pi\varepsilon_0}\,\sqrt{8c}\,\sqrt{2c}\,e^2 A_r(e)}{c \, m_e\lambda_p} = \frac{\sqrt{\pi\varepsilon_0}\,\sqrt{8c}\,\sqrt{2c}\,e^2 R}{c \, K_B\lambda_p} = \frac{\sqrt{\pi\varepsilon_0}\,\sqrt{8c}\,\sqrt{2c}\,e^2 N_A}{c \, A_r(e)\lambda_e} = \frac{\sqrt{\pi\varepsilon_0}\,\sqrt{8c}\,\sqrt{2c}\,e^2}{m_e\lambda_e \, c}$$

$$= \frac{\sqrt{\pi\varepsilon_0}\,\sqrt{8c}\,\sqrt{2c}\,e^2 N_A}{c \, \lambda_p} = \frac{\sqrt{8c}\,\sqrt{2c}\,\alpha^2\hbar\varepsilon_0}{e^2 c} = \frac{\alpha\sqrt{\varepsilon_0\mu_0}}{\sqrt{\pi\varepsilon_0}\,\sqrt{8c}\,\sqrt{2c}\,\pi e^2} = \frac{\sqrt{8c}\,\sqrt{2c}\,e^2 c\varepsilon_0}{h}$$

Speed of light:

$$c = \frac{\alpha \hbar}{e^2} = \frac{2h\alpha}{e^2\mu_0} = \frac{1}{\sqrt{\varepsilon_0\,\mu_0}} = \frac{1}{c\,\varepsilon_0\,\mu_0} = \textbf{299,792,458 meters per second}$$

$$\frac{1}{c} = \textbf{3.335,640,951,981,520,495,755,767,144,749,2e-9 meters per second}$$

$$\frac{1}{c} = \frac{e^2}{\alpha\,\hbar} = \frac{e^2\mu_0}{2h\alpha} = \sqrt{\varepsilon_0\,\mu_0} = c\,\varepsilon_0\,\mu_0$$

$$c^2 = \frac{\alpha^2\,\hbar^2}{e^4} = \frac{1}{\varepsilon_0\,\mu_0} = \textbf{89,875,517,873,681,764 meters per second}$$

$$\frac{1}{c^2} = \frac{e^4}{\alpha^2\,\hbar^2} = c^2(\varepsilon_0\mu_0)^2 = \varepsilon_0\,\mu_0 = \textbf{1.112,650,056,053,618,432,174,089,964,848e-17 m/s}$$

$$c^3 = \frac{\alpha^3\,\hbar^3}{e^6} = \frac{c}{\varepsilon_0\,\mu_0} = 26,944,002,417,373,989,539,335,912 \text{ m/s}$$

$$\frac{1}{c^3} = \frac{e^6}{\alpha^3\,\hbar^3} = \frac{\varepsilon_0\,\mu_0}{c} = 3.711,401,092,196,983,928,708,740,113,962,5e-26 \text{ m/s}$$

$$c^4 = \frac{\alpha^4\,\hbar^4}{e^8} = \frac{1}{(\varepsilon_0\,\mu_0)^2} = 807,760,871,306,249,022,926,338,007,461,517 \text{ m/s}$$

$$\frac{1}{c^4} = \frac{e^8}{\alpha^4\hbar^4} = (\varepsilon_0\,\mu_0)^2 = 1.237,990,147,236,120,239,125,141,738,543,2e-34 \text{ m/s}$$

$$c^5 = \frac{\alpha^5\hbar^5}{e^{10}} = \frac{c}{(\varepsilon_0\,\mu_0)^2} = 24,216,061,708,512,206,534,319,783,561,111e42 \text{ m/s}$$

$$\frac{1}{c^5} = \frac{e^{10}}{\alpha^5\hbar^5} = \frac{(\varepsilon_0\,\mu_0)^2}{c} = 4.129,490,633,270,434,839,041,687,094,553,8e-43 \text{ m/s}$$

$$c^6 = \frac{\alpha^6\hbar^6}{e^{12}} = \frac{1}{(\varepsilon_0\mu_0)^3} = \frac{c^2}{(\varepsilon_0\mu_0)^2} = 72,597,926,626,745,539,199,273,892,718,133e50 \text{ m/s}$$

$$\frac{1}{c^6} = \frac{e^{12}}{\alpha^6\hbar^6} = (\varepsilon_0\,\mu_0)^3 = \frac{(\varepsilon_0\mu_0)^2}{c^2} = 1.377,449,806,716,096,520,027,093,908,597e-51 \text{ m/s}$$

$$c^7 = \frac{\alpha^7 h^7}{e^{14}} = \frac{c}{(\varepsilon_0 \mu_0)^3} = 21{,}764{,}310{,}869{,}135{,}693{,}737{,}085{,}672{,}113{,}197e59 \text{ m/s}$$

$$\frac{1}{c^7} = \frac{e^{14}}{\alpha^7 h^7} = \frac{(\varepsilon_0 \mu_0)^3}{c} = 4.594{,}677{,}984{,}581{,}241{,}600{,}237{,}634{,}759{,}303{,}6e\text{-}60 \text{ m/s}$$

$$c^8 = \frac{\alpha^8 h^8}{e^{16}} = \frac{1}{(\varepsilon_0\mu_0)^4} = \frac{c^2}{(\varepsilon_0\mu_0)^3} = 65{,}247{,}762{,}521{,}343{,}059{,}609{,}761{,}193{,}993{,}975e36 \text{ m/s}$$

$$\frac{1}{c^8} = \frac{e^{16}}{\alpha^8 h^8} = (\varepsilon_0 \mu_0)^4 = \frac{(\varepsilon_0\mu_0)^3}{c^2} = 1.532{,}619{,}604{,}653{,}710{,}668{,}144{,}171{,}511{,}914e\text{-}36 \text{ m/s}$$

$$c^9 = \frac{\alpha^9 h^9}{e^{18}} = \frac{c}{(\varepsilon_0 \mu_0)^4} = 19{,}560{,}787{,}105{,}273{,}713{,}301{,}650{,}829{,}140{,}469e45 \text{ m/s}$$

$$\frac{1}{c^9} = \frac{e^{18}}{\alpha^9 h^9} = \frac{(\varepsilon_0 \mu_0)^4}{c} = 5.112{,}268{,}717{,}092{,}645{,}032{,}931{,}987{,}608{,}288{,}3e\text{-}77 \text{ m/s}$$

$$c^{10} = \frac{\alpha^{10} h^{10}}{e^{20}} = \frac{1}{(\varepsilon_0\mu_0)^5} = \frac{c^2}{(\varepsilon_0\mu_0)^4} = 5.864{,}176{,}446{,}704{,}711{,}273{,}489{,}197{,}525{,}75e84 \text{ m/s}$$

$$\frac{1}{c^{10}} = \frac{e^{20}}{\alpha^{10}h^{10}} = (\varepsilon_0 \mu_0)^5 = \frac{(\varepsilon_0\mu_0)^4}{c^2} = 1.705{,}269{,}289{,}026{,}825{,}695{,}839{,}215{,}410{,}9e\text{-}85 \text{ m/s}$$

Elementary charge:

$$e = \sqrt{\frac{\alpha h}{c}} = 1.602{,}176{,}564{,}374{,}559{,}619{,}029{,}512{,}522{,}979{,}9e\text{-}19 \text{ Coulomb}$$

$$e^2 = \frac{\alpha h}{c} = \frac{2h\alpha}{c \mu_0} = 2.566{,}969{,}743{,}431{,}067{,}383{,}000{,}001{,}685{,}547{,}5e\text{-}38 \text{ Coulomb}$$

$$e^2 = \alpha \hbar c \varepsilon_0 \mu_0 = \frac{m_e \lambda_e \alpha}{\sqrt{\pi\varepsilon_0} \sqrt{8c} \sqrt{2c} \pi} = \frac{m_u \lambda_p \alpha}{\sqrt{\pi\varepsilon_0} \sqrt{8c} \sqrt{2c} \pi} = \frac{\alpha A_r(e)\lambda_e}{\sqrt{\pi\varepsilon_0} \sqrt{8c} \sqrt{2c} \pi N_A}$$

$$= \alpha \hbar \sqrt{\varepsilon_0\mu_0} = \frac{\alpha \lambda_p}{\sqrt{\pi\varepsilon_0} \sqrt{8c} \sqrt{2c} \pi N_A} = \frac{\alpha m_e\lambda_p}{\sqrt{\pi\varepsilon_0} \sqrt{8c} \sqrt{2c} \pi A_r(e)} = \frac{\alpha K_B\lambda_p}{\sqrt{\pi\varepsilon_0} \sqrt{8c} \sqrt{2c} \pi R}$$

$$= \frac{\alpha h}{\sqrt{8c} \sqrt{2c} \pi c_0 \varepsilon_0}$$

Fine Structure Constant α:

$$\alpha = \frac{c\,e^2}{h} = \frac{c\mu_0}{2R_K} = \frac{ce^2\mu_0}{2h} = 7.297,352,569,800,184,997,640,004,791,553,6e\text{-}3$$

$$= \frac{e^2}{h\sqrt{\varepsilon_0\mu_0}} = \frac{e^2}{h\,c\,\varepsilon_0\,\mu_0} = \frac{\sqrt{\pi\varepsilon_0}\,\sqrt{8c}\,\sqrt{2c}\,\pi e^2}{m_e\lambda_e} = \frac{\sqrt{\pi\varepsilon_0}\,\sqrt{8c}\,\sqrt{2c}\,\pi e^2}{m_u\lambda_p} = \frac{\sqrt{\pi\varepsilon_0}\,\sqrt{8c}\,\sqrt{2c}\,\pi e^2\,N_A}{A_r(e)\lambda_e}$$

$$= \frac{\sqrt{\pi\varepsilon_0}\,\sqrt{8c}\,\sqrt{2c}\,\pi e^2\,N_A}{\lambda_p} = \frac{\sqrt{\pi\varepsilon_0}\,\sqrt{8c}\,\sqrt{2c}\,\pi e^2 R}{K_B\lambda_p} = \frac{\sqrt{\pi\varepsilon_0}\,\sqrt{8c}\,\sqrt{2c}\,\pi e^2\,A_r(e)}{c\,m_e\lambda_p}$$

Planck Constant h:

$$h = 2\pi\hbar = \frac{\pi\hbar}{5} = 6.626,069,570,023,165,074,017,466,569,367,9e\text{-}34 \; J\cdot S$$

$$= m_e\lambda_e\,c = m_u\lambda_p\,c = \sqrt{\pi\varepsilon_0}\,\sqrt{8c}\,\sqrt{2c}\,\pi\hbar = \frac{c\,\lambda_p}{N_A} = \frac{c\,m_e\lambda_p}{A_r(e)} = \frac{c\,K_B\lambda_p}{R} = \frac{c\,A_r(e)\lambda_e}{N_A}$$

$$= \frac{\pi\hbar}{\sqrt{\pi\varepsilon_0}\,c} = \frac{\sqrt{\pi\varepsilon_0}\,e^2}{\alpha\varepsilon_0} = \frac{\pi e^2}{\sqrt{\pi\varepsilon_0}\,\alpha} = \frac{\sqrt{\pi\varepsilon_0}\,\hbar}{c\varepsilon_0} = \frac{\sqrt{\pi\varepsilon_0}\,e^4 c}{\alpha^2\hbar\varepsilon_0} = \frac{\sqrt{\pi\varepsilon_0}\,\sqrt{8c}\,\sqrt{2c}\,\pi e^2 c}{\alpha}$$

$$= \frac{\sqrt{\pi\varepsilon_0}\,\sqrt{8c}\,\sqrt{2c}\,\pi e^2}{\alpha\sqrt{\varepsilon_0\mu_0}} = \frac{\sqrt{\pi\varepsilon_0}\,\sqrt{8c}\,\sqrt{2c}\,\pi e^2}{\alpha\,c\,\varepsilon_0\,\mu_0} = \frac{e^2}{\sqrt{\pi\varepsilon_0}\,\sqrt{8c}\,\sqrt{2c}\,\alpha\,c\,\varepsilon_0}$$

Reduced Planck Constant ℏ:

$$\hbar = 1.054,571,725,339,976,234,000,102,692,456,8e\text{-}34 \; J\cdot S$$

$$\hbar = \frac{h}{2\pi} = \frac{e^2 c}{\alpha} = \frac{2h}{\mu_0} = \frac{e^2}{\alpha\sqrt{\varepsilon_0\mu_0}} = \frac{e^2}{\alpha\,c\,\varepsilon_0\,\mu_0} = \frac{m_e\lambda_e\,c}{\sqrt{\pi\varepsilon_0}\,\sqrt{8c}\,\sqrt{2c}\,\pi} = \frac{m_u\lambda_p\,c}{\sqrt{\pi\varepsilon_0}\,\sqrt{8c}\,\sqrt{2c}\,\pi}$$

$$= \frac{c\,\lambda_p}{\sqrt{\pi\varepsilon_0}\,\sqrt{8c}\,\sqrt{2c}\,\pi N_A} = \frac{c\,m_e\lambda_p}{\sqrt{\pi\varepsilon_0}\,\sqrt{8c}\,\sqrt{2c}\,\pi A_r(e)} = \frac{c\,K_B\lambda_p}{\sqrt{\pi\varepsilon_0}\,\sqrt{8c}\,\sqrt{2c}\,\pi R} = \frac{c\,A_r(e)\lambda_e}{\sqrt{\pi\varepsilon_0}\,\sqrt{8c}\,\sqrt{2c}\,\pi N_A}$$

$$= \frac{\hbar}{\sqrt{8c}\,\sqrt{2c}\,\pi c\varepsilon_0} = \frac{e^2}{\sqrt{8c}\,\sqrt{2c}\,\pi\alpha\varepsilon_0} = \frac{e^4 c}{\sqrt{8c}\,\sqrt{2c}\,\pi\alpha^2\hbar\varepsilon_0}$$

Numerical constants:

$$\pi = \frac{e^4}{4\alpha^2 h^2 \varepsilon_0} = \frac{1}{4c^2\varepsilon_0} = \frac{h}{2\hbar} = \frac{\alpha h}{2ce^2} = 3.141,592,653,589,793,238,462,643,383,279,5$$

$$4\pi = \mu_0 = \frac{e^4}{\alpha^2 h^2 \varepsilon_0} = \frac{1}{c^2\varepsilon_0} = \frac{2h}{h} = \frac{2\alpha h}{ce^2} = 12.566,370,614,359,172,953,850,573,533,118$$

$$8\pi = 2\mu_0 = \frac{2e^4}{\alpha^2 h^2 \varepsilon_0} = \frac{2}{c^2\varepsilon_0} = \frac{4h}{h} = \frac{4\alpha h}{ce^2} = 25.132,741,228,718,345,907,701,147,066,241$$

$$8\pi c^2 = \frac{2c^2 e^4}{\alpha^2 h^2 \varepsilon_0} = \frac{2}{\varepsilon_0} = \frac{4hc^2}{\hbar} = \frac{4\alpha hc}{e^2} = 2.258,818,133,516,294,276,701,646,785,873,5$$

$$8c\pi^2 = 2c\,4\pi^2 = 2.367,066,370,312,157,358,387,101,721,625,4$$

$$8\pi\varepsilon_0 = \frac{2}{c^2} = 2\varepsilon_0\mu_0 = 2.225,300,112,107,236,864,348,179,929,696$$

$$\frac{\sqrt{2c}}{2} = \frac{2c}{\sqrt{8c}} = 1.224,321,154,762,915,954,410,585,658,518,2$$

$$\sqrt{2c} = \frac{4c}{\sqrt{8c}} = 2.448,642,309,525,831,908,821,171,317,036,5$$

$$\sqrt{2c}\,2\pi = \sqrt{8c}\,\pi = 1.538,527,338,175,099,623,207,488,901,824,8$$

$$\sqrt{2c}\,4\pi = \sqrt{8c}\,2\pi = 3.077,054,676,350,199,246,414,977,803,649,5$$

$$\sqrt{2\pi} = \frac{\sqrt{8\pi}}{2} = 2.506,628,274,631,000,502,415,765,284,811$$

$$\sqrt{8\pi} = \sqrt{2\pi}\,2 = 5.013,256,549,262,001,004,831,530,569,622,6$$

$$1 \quad = \sqrt{8\pi}\ \sqrt{8\varepsilon_0}\ \pi c^3 \varepsilon_0 = \frac{\pi c \varepsilon_0}{(\sqrt{\pi\varepsilon_0}\ -\ \pi c \varepsilon_0)} = \frac{\sqrt{\pi\varepsilon_0}\ \sqrt{8c}\ c}{\sqrt{2c}} = \frac{(\pi c^3 \varepsilon_0 - c)}{(\sqrt{\pi\varepsilon_0} + \pi c \varepsilon_0)\sqrt{\pi\varepsilon_0}\ c^3}$$

$$2 \quad = \sqrt{\pi\varepsilon_0}\ \sqrt{8c}\ \sqrt{2c}\ = \frac{1}{\sqrt{\pi\varepsilon_0}\ c} = \frac{\sqrt{\pi\varepsilon_0}}{\pi c \varepsilon_0} = \frac{\alpha^2 \hbar^2}{\sqrt{\pi\varepsilon_0}\ e^4 c^3}$$

$$3 \quad = (\sqrt{\pi\varepsilon_0} + \pi c \varepsilon_0)\ c\ = \frac{(\pi c^3 \varepsilon_0 - c)}{\sqrt{\pi\varepsilon_0}\ c^2}$$

$$4 \quad = \frac{1}{\pi c^2 \varepsilon_0} = \frac{e^4}{\pi \alpha^2 \hbar^2 \varepsilon_0} = \frac{(\pi c \varepsilon_0 - \sqrt{\pi\varepsilon_0})}{\sqrt{\pi\varepsilon_0}}$$

$$5 \quad = \sqrt{\pi\varepsilon_0}\ c\ = \frac{\pi c \varepsilon_0}{\sqrt{\pi\varepsilon_0}} = \frac{\sqrt{2c}}{\sqrt{8c}} = \frac{1}{\sqrt{\pi\varepsilon_0}\ \sqrt{8c}\ \sqrt{2c}}$$

$$6 \quad = \frac{(\sqrt{\pi\varepsilon_0} + \pi c \varepsilon_0)}{\sqrt{\pi\varepsilon_0}} = (\sqrt{\pi\varepsilon_0} + \pi c \varepsilon_0)\sqrt{\pi\varepsilon_0}\ \sqrt{8c}\ \sqrt{2c}\ c$$

$$7 \quad = \frac{1}{\sqrt{\pi\varepsilon_0}\ c} + \sqrt{\pi\varepsilon_0}\ c$$

$$8 \quad = \frac{1}{\sqrt{\pi\varepsilon_0}\ \pi c^3 \varepsilon_0} = \frac{\sqrt{\pi\varepsilon_0}}{\pi^2 c^3 \varepsilon_0^2} = \frac{\sqrt{\pi\varepsilon_0}\ \sqrt{8c}\ \sqrt{2c}}{\pi c^2 \varepsilon_0}$$

$$9 \quad = (\sqrt{\pi\varepsilon_0} + \pi c \varepsilon_0)^2\ c^2$$

$$10 \quad = \sqrt{8c}\ \sqrt{2c}\ \pi c \varepsilon_0\ = \left(\pi c \varepsilon_0\ -\ \sqrt{\pi\varepsilon_0}\right)\sqrt{\pi\varepsilon_0}\ c$$

$$11 \quad = \frac{(\sqrt{\pi\varepsilon_0} + \pi c \varepsilon_0)}{\sqrt{\pi\varepsilon_0}} + \sqrt{\pi\varepsilon_0}\ c$$

$$12 \quad = \frac{(\sqrt{\pi\varepsilon_0} + \pi c \varepsilon_0)}{\pi c \varepsilon_0} = (\sqrt{\pi\varepsilon_0} + \pi c \varepsilon_0)\sqrt{8c}\ \sqrt{2c}$$

$$13 \quad = \frac{\sqrt{\pi\varepsilon_0}}{\pi^2 c^3 \varepsilon_0^2} + \sqrt{\pi\varepsilon_0}\ c$$

$$14 \quad = \left(\frac{1}{\sqrt{\pi\varepsilon_0}\ c} + \sqrt{\pi\varepsilon_0}\ c\right)\frac{1}{\sqrt{\pi\varepsilon_0}\ c}$$

$$15 \quad = (\sqrt{\pi\varepsilon_0} + \pi c \varepsilon_0)\sqrt{\pi\varepsilon_0}\ c^2\ = \frac{(\pi c^3 \varepsilon_0 - c)}{c}$$

$$16 \quad = \quad \frac{1}{\pi^2 c^4 \varepsilon_0^2}$$

$$18 \quad = \quad \frac{(\sqrt{\pi \varepsilon_0} + \pi c \varepsilon_0)^2 c}{\sqrt{\pi \varepsilon_0}}$$

$$24 \quad = \quad \frac{(\sqrt{\pi \varepsilon_0} + \pi c \varepsilon_0)}{\sqrt{\pi \varepsilon_0} \, \pi c^2 \varepsilon_0}$$

$$27 \quad = \quad (\sqrt{\pi \varepsilon_0} + \pi c \varepsilon_0)^3 c^3$$

$$25 \quad = \quad \pi c^2 \varepsilon_0 \quad = \quad \frac{\pi a^2 h^2 \varepsilon_0}{e^4} \quad = \quad \frac{\sqrt{\pi \varepsilon_0}}{(\pi c \varepsilon_0 - \sqrt{\pi \varepsilon_0})}$$

$$32 \quad = \quad \frac{\sqrt{\pi \varepsilon_0}}{\pi^3 c^5 \varepsilon_0^3} \quad = \quad \frac{1}{\sqrt{\pi \varepsilon_0} \, \pi^2 c^5 \varepsilon_0^2}$$

$$45 \quad = \quad (\sqrt{\pi \varepsilon_0} + \pi c \varepsilon_0)^2 \sqrt{\pi \varepsilon_0} \, c^3 \quad = \quad (\sqrt{\pi \varepsilon_0} + \pi c \varepsilon_0)(\pi c^3 \varepsilon_0 - c)$$

$$64 \quad = \quad \frac{1}{\pi^3 c^6 \varepsilon_0^3}$$

$$72 \quad = \quad \frac{(\sqrt{\pi \varepsilon_0} + \pi c \varepsilon_0)^2 \sqrt{\pi \varepsilon_0}}{\pi^2 c \varepsilon_0^2}$$

$$75 \quad = \quad (\sqrt{\pi \varepsilon_0} + \pi c \varepsilon_0)\pi c^3 \varepsilon_0 \quad = \quad (\pi c^3 \varepsilon_0 - c)\sqrt{\pi \varepsilon_0}$$

$$81 \quad = \quad (\sqrt{\pi \varepsilon_0} + \pi c \varepsilon_0)^4 c^4$$

$$96 \quad = \quad \frac{(\sqrt{\pi \varepsilon_0} + \pi c \varepsilon_0)}{\sqrt{\pi \varepsilon_0} \, \pi^2 c^4 \varepsilon_0^2} \quad = \quad \frac{(\sqrt{\pi \varepsilon_0} + \pi c \varepsilon_0)\sqrt{8c} \, \sqrt{2c}}{\sqrt{\pi \varepsilon_0} \, \pi c^3 \varepsilon_0} \quad = \quad \frac{(\sqrt{\pi \varepsilon_0} + \pi c \varepsilon_0)\sqrt{\pi \varepsilon_0}}{\pi^3 c^4 \varepsilon_0^3}$$

$$= \quad \frac{(\sqrt{\pi \varepsilon_0} + \pi c \varepsilon_0)\sqrt{\pi \varepsilon_0} \, \sqrt{8c} \, \sqrt{2c}}{\pi^2 c^3 \varepsilon_0^2}$$

$$100 \quad = \quad (\pi c \varepsilon_0 - \sqrt{\pi \varepsilon_0})\sqrt{\pi \varepsilon_0} \, c^2 \quad = \quad \sqrt{8c} \, \sqrt{2c} \, \pi c \varepsilon_0 \quad (10)$$

$$125 \quad = \quad \sqrt{\pi \varepsilon_0} \, \pi c^3 \varepsilon_0$$

$$128 = \frac{\sqrt{\pi\varepsilon_0}}{\pi^4 c^7 \varepsilon_0^4}$$

$$375 = (\sqrt{\pi\varepsilon_0} + \pi c\varepsilon_0)\sqrt{\pi\varepsilon_0}\,\pi c^4\varepsilon_0 = \frac{(\pi c^3\varepsilon_0 - c)}{\sqrt{8c}\,\sqrt{2c}}$$

$$512 = \frac{\sqrt{\pi\varepsilon_0}}{\pi^3 c^5 \varepsilon_0^3}$$

$$625 = \pi^2 c^4 \varepsilon_0^2$$

$$768 = \frac{(\sqrt{\pi\varepsilon_0} + \pi c\varepsilon_0)}{\pi^4 c^7 \varepsilon_0^4}$$

$$1024 = \frac{1}{\pi^5 c^{10} \varepsilon_0^5}$$

$$599{,}584{,}916 = \frac{1}{\sqrt{\pi\varepsilon_0}} = \frac{\sqrt{\pi\varepsilon_0}}{\pi\varepsilon_0} = 2c$$

$$1{,}599{,}584{,}916 = \frac{(\sqrt{\pi\varepsilon_0} + \pi\varepsilon_0)}{\pi\varepsilon_0}$$

$$1.049{,}273{,}603 = (\pi c^3 \varepsilon_0 + c)$$

Coulomb constant: $k_e = \dfrac{\alpha^2 \hbar^2}{e^4} = \dfrac{1}{\varepsilon_0\,\mu_0} = 8.987{,}551{,}787{,}368{,}176{,}4\text{E}9\ \text{N·m}^2/\text{C}^2$

Electromagnetic coupling constant:

$$g = \frac{e}{\sqrt{\hbar c}} = \frac{\sqrt{\alpha}}{c} = \sqrt{\alpha\,\varepsilon_0\mu_0} = 2.849{,}456{,}043{,}849{,}631{,}373{,}051{,}310{,}841{,}799$$

Von Klitzing constant:

$$R_K \;=\; \frac{h}{e^2} \;=\; \frac{c\mu_0}{2\alpha} \;=\; 25{,}812.807{,}443{,}404{,}521{,}417{,}322{,}882{,}144{,}935 \text{ Ohms}$$

$$R_K \;=\; \frac{m_e \lambda_e\, c}{e^2} \;=\; \frac{m_u \lambda_p\, c}{e^2} \;=\; \frac{c\,\lambda_p}{e^2\, N_A} \;=\; \frac{c\, m_e \lambda_p}{e^2\, A_r(e)} \;=\; \frac{c\, K_B \lambda_p}{e^2\, R} \;=\; \frac{c\, A_r(e)\lambda_e}{e^2\, N_A} \;=\; \frac{\sqrt{\pi\varepsilon_0}}{\alpha\varepsilon_0}$$

$$=\; \frac{\pi}{\sqrt{\pi\varepsilon_0}\,\alpha} \;=\; \frac{\sqrt{\pi\varepsilon_0}\, h}{e^2 c \varepsilon_0} \;=\; \frac{\sqrt{\pi\varepsilon_0}\, e^2 c}{\alpha^2 \hbar \varepsilon_0} \;=\; \frac{\pi \hbar}{\sqrt{\pi\varepsilon_0}\, e^2 c} \;=\; \frac{\sqrt{\pi\varepsilon_0}\,\sqrt{8c}\,\sqrt{2c}\,\pi c}{\alpha} \;=\; \frac{\sqrt{\pi\varepsilon_0}\,\sqrt{8c}\,\sqrt{2c}\,\pi}{\alpha c \varepsilon_0 \mu_0}$$

$$=\; \frac{\sqrt{\pi\varepsilon_0}\, e^2 c^3 \mu_0}{\alpha^2 h} \;=\; \frac{\sqrt{\pi\varepsilon_0}\,\sqrt{8c}\,\sqrt{2c}\,\pi}{\alpha\sqrt{\varepsilon_0 \mu_0}} \;=\; \frac{\sqrt{\pi\varepsilon_0}\,\sqrt{8c}\,\sqrt{2c}\,\pi h}{e^2} \;=\; \frac{1}{\sqrt{\pi\varepsilon_0}\,\sqrt{8c}\,\sqrt{2c}\,\alpha\, c\, \varepsilon_0}$$

Hall effect: $\;=\; \dfrac{e^2}{h} \;=\; \dfrac{2\alpha}{c\mu_0} \;=\; .387{,}404{,}586{,}731{,}813{,}189{,}681{,}564{,}599{,}955{,}21$

$$=\; \frac{8\pi\alpha^2 \hbar \varepsilon_0}{e^2 \mu_0} \;=\; \frac{\alpha\sqrt{\varepsilon_0 \mu_0}}{2\pi} \;=\; \frac{\alpha\, c\, \varepsilon_0\, \mu_0}{2\pi} \;=\; \frac{e^2}{m_e \lambda_e\, c} \;=\; \frac{e^2}{m_u \lambda_p\, c} \;=\; \frac{e^2\, N_A}{c\,\lambda_p} \;=\; \frac{e^2\, A_r(e)}{c\, m_e \lambda_p} \;=\; \frac{e^2\, R}{c\, K_B \lambda_p}$$

$$=\; \frac{e^2\, N_A}{c\, A_r(e)\lambda_e} \;=\; \frac{\alpha\varepsilon_0}{\sqrt{\pi\varepsilon_0}} \;=\; \frac{\sqrt{\pi\varepsilon_0}\,\alpha}{\pi} \;=\; \frac{e^2 c \varepsilon_0}{\sqrt{\pi\varepsilon_0}\, h} \;=\; \frac{\alpha^2 \hbar \varepsilon_0}{\sqrt{\pi\varepsilon_0}\, e^2 c} \;=\; \frac{\alpha^2 h}{\sqrt{\pi\varepsilon_0}\, e^2 c^3 \mu_0} \;=\; \frac{\alpha}{\sqrt{\pi\varepsilon_0}\,\sqrt{8c}\,\sqrt{2c}\,\pi c}$$

$$=\; \frac{\alpha c \varepsilon_0 \mu_0}{\sqrt{\pi\varepsilon_0}\,\sqrt{8c}\,\sqrt{2c}\,\pi} \;=\; \frac{\sqrt{\pi\varepsilon_0}\, e^2 c}{\pi h} \;=\; \frac{\alpha\sqrt{\varepsilon_0 \mu_0}}{\sqrt{\pi\varepsilon_0}\,\sqrt{8c}\,\sqrt{2c}\,\pi} \;=\; \frac{e^2}{\sqrt{\pi\varepsilon_0}\,\sqrt{8c}\,\sqrt{2c}\,\pi h} \;=\; \sqrt{\pi\varepsilon_0}\,\sqrt{8c}\,\sqrt{2c}\,\alpha c \varepsilon_0$$

Josephson constant:

$$K_J \;=\; \frac{2e}{h} \;=\; \frac{e}{\pi h} \;=\; \frac{2\sqrt{\alpha\, \hbar}}{h\sqrt{c}} \;=\; 483{,}597.869{,}730{,}473{,}816{,}563{,}327{,}973{,}661{,}2 \text{ Hz/Volt}$$

$$=\; \frac{\alpha}{\pi e c} \;=\; \frac{\alpha c \varepsilon_0 \mu_0}{\pi e} \;=\; \frac{\alpha\sqrt{\varepsilon_0 \mu_0}}{\pi e} \;=\; \frac{\sqrt{\pi\varepsilon_0}\,\sqrt{8c}\,\sqrt{2c}\, e}{h} \;=\; \frac{\sqrt{\pi\varepsilon_0}\,\sqrt{8c}\,\sqrt{2c}\, e}{c\, m_e \lambda_e} \;=\; \frac{\sqrt{\pi\varepsilon_0}\,\sqrt{8c}\,\sqrt{2c}\, e}{c\, m_u \lambda_p}$$

$$=\; \frac{\sqrt{\pi\varepsilon_0}\,\sqrt{8c}\,\sqrt{2c}\, e\, A_r(e)}{c\, m_e \lambda_p} \;=\; \frac{\sqrt{\pi\varepsilon_0}\,\sqrt{8c}\,\sqrt{2c}\, e\, R}{c\, K_B \lambda_p} \;=\; \frac{\sqrt{8c}\,\sqrt{2c}\, e \varepsilon_0\, N_A}{h\, A_r(e)\lambda_e} \;=\; \frac{\sqrt{8c}\,\sqrt{2c}\,\alpha\varepsilon_0}{e}$$

$$=\; \frac{\sqrt{8c}\,\sqrt{2c}\, e c \varepsilon_0}{h} \;=\; \frac{\sqrt{8c}\,\sqrt{2c}\,\alpha^2 \hbar \varepsilon_0}{e^3 c} \;=\; \frac{\sqrt{\pi\varepsilon_0}\,\sqrt{8c}\,\sqrt{2c}\, e\, N_A}{c\,\lambda_p}$$

$$\Phi_0 = \frac{h}{2e} = \frac{\pi\hbar}{e} = \frac{\pi e c}{\alpha} = \frac{h\sqrt{c}}{2\sqrt{\alpha\,\hbar}} = 2.067{,}833{,}757{,}326{,}795{,}795{,}427{,}727{,}403{,}567{,}2$$

$$= \frac{\pi e}{\alpha c \varepsilon_0 \mu_0} = \frac{\pi e}{\alpha\sqrt{\varepsilon_0 \mu_0}} = \frac{e}{\sqrt{8c}\,\sqrt{2c}\,\alpha\varepsilon_0} = \frac{\hbar}{\sqrt{8c}\,\sqrt{2c}\,ec\varepsilon_0} = \frac{c\,m_e\lambda_e}{\sqrt{\pi\varepsilon_0}\,\sqrt{8c}\,\sqrt{2c}\,e} = \frac{c\,m_u\lambda_p}{\sqrt{\pi\varepsilon_0}\,\sqrt{8c}\,\sqrt{2c}\,e}$$

$$= \frac{h}{\sqrt{\pi\varepsilon_0}\,\sqrt{8c}\,\sqrt{2c}\,e} = \frac{c\,m_e\lambda_p}{\sqrt{\pi\varepsilon_0}\,\sqrt{8c}\,\sqrt{2c}\,e\,A_r(e)} = \frac{c\,K_B\lambda_p}{\sqrt{\pi\varepsilon_0}\,\sqrt{8c}\,\sqrt{2c}\,e\,R} = \frac{\hbar\,A_r(e)\lambda_e}{\sqrt{8c}\,\sqrt{2c}\,e\varepsilon_0\,N_A}$$

$$= \frac{e^3 c}{\sqrt{8c}\,\sqrt{2c}\,\alpha^2\hbar\varepsilon_0} = \frac{c\,\lambda_p}{\sqrt{\pi\varepsilon_0}\,\sqrt{8c}\,\sqrt{2c}\,e\,N_A}$$

Gravitational constant:

$$G \quad = \quad 6.672{,}536{,}659{,}401{,}633{,}510{,}465{,}095{,}670{,}374{,}1e\text{-}11 \ \text{m}^3\text{Kg}^{-1}\text{sec}^{-2}$$

$$G = \frac{\hbar c}{8\pi e^{16}} = \frac{c^5}{8\pi e^8\alpha^4\hbar^3} = \frac{\pi\hbar c^4\varepsilon_0^2}{\sqrt{\pi\varepsilon_0}\,e^{16}} = \frac{\sqrt{\pi\varepsilon_0}\,\hbar c^4\varepsilon_0}{e^{16}} = \frac{\pi c^8\varepsilon_0^2}{\sqrt{\pi\varepsilon_0}\,e^8\alpha^4\hbar^3} = \frac{\sqrt{\pi\varepsilon_0}\,c^8\varepsilon_0}{e^8\alpha^4\hbar^3}$$

Gravitational coupling constant:

$$\alpha_G \quad = \quad 1.751{,}314{,}084{,}598{,}958{,}167{,}761{,}396{,}803{,}218{,}2e\text{-}45$$

$$\alpha_G \quad = \quad \frac{G m_e^2}{\hbar c} \quad = \quad \frac{\pi\alpha^4\hbar^6 c^3}{e^{16}} \quad = \quad \frac{\pi c^2}{2\lambda_e^2 e^8\alpha^4\hbar^2} \quad = \quad \frac{m_e^2}{8\pi e^{16}}$$

Boltzmann constant:

$$K_B = R\,m_u = \frac{R}{N_A} = 1.380{,}648{,}798{,}394{,}132{,}924{,}292{,}774{,}930{,}237{,}3e\text{-}23 \ \text{J/K}$$

$$= \frac{R\,m_e}{A_r(e)} = \frac{h\,R}{c\,\lambda_p} = \frac{A_r(e)\,\lambda_e\,R}{N_A\,\lambda_p} = \frac{m_e\lambda_e R}{\lambda_p} = \frac{\sqrt{\pi\varepsilon_0}\,h\,R}{c^2\varepsilon_0\,\lambda_p} = \frac{\pi\hbar\,R}{\sqrt{\pi\varepsilon_0}\,c^2\,\lambda_p} = \frac{\sqrt{\pi\varepsilon_0}\,e^2\,R}{\alpha c\varepsilon_0\,\lambda_p}$$

$$= \frac{\sqrt{\pi\varepsilon_0}\,e^4\,R}{\alpha^2\hbar\varepsilon_0\,\lambda_p} = \frac{\pi e^2\,R}{\sqrt{\pi\varepsilon_0}\,\alpha c\,\lambda_p} = \frac{4\pi\hbar R_\infty R}{A_r(e)\,\alpha^2 c}$$

Stefan-Boltzmann constant:

$$\sigma \qquad = \qquad 5.670,257,126,259,703,311,750,510,614,457,1\text{e-}8 \ W_m$$

$$\sigma \quad = \quad \frac{\pi^2 K_B^4}{60 h^3 c^2} \quad = \quad \frac{2\pi^5 K_B^4}{15 c^2 h^3} \quad = \quad \frac{\pi^2 c^2}{60 \alpha^4 h^3 \varepsilon_0^2} \quad = \quad \frac{2\pi^5 c^2}{15 \alpha^4 h^3 \varepsilon_0^2}$$

Mass of an electron:

$$m_e = m_u A_r(e) = \frac{e^2}{r_e} = \frac{h}{c \lambda_e} = 9.109,382,903,345,524,937,498,006,139,342,6\text{e-}31 \ Kg$$

$$= \frac{m_u \lambda_p}{\lambda_e} = \frac{\hbar R_\infty}{\alpha^2 c^3 \varepsilon_0} = \frac{h A_r(e)}{c \lambda_p} = \frac{A_r(e)}{N_A} = \frac{A_r(e) K_B}{R} = \frac{h \mu_0 R_\infty}{\alpha^2 c} = \frac{\sqrt{\varepsilon_0 \mu_0} \, 4\pi h R_\infty}{\alpha^2} = \frac{\sqrt{\pi \varepsilon_0} \, h A_r(e)}{c^2 \varepsilon_0 \lambda_p}$$

$$= \frac{\pi h A_r(e)}{\sqrt{\pi \varepsilon_0} \, c^2 \lambda_p} = \frac{\sqrt{\pi \varepsilon_0} \, e^2 A_r(e)}{\alpha c \varepsilon_0 \lambda_p} = \frac{\pi e^2 A_r(e)}{\sqrt{\pi \varepsilon_0} \, \alpha c \lambda_p} = \frac{\sqrt{\pi \varepsilon_0} \, h}{c^2 \varepsilon_0 \lambda_e} = \frac{\pi h}{\sqrt{\pi \varepsilon_0} \, c^2 \lambda_e} = \frac{\sqrt{\pi \varepsilon_0} \, e^2}{\alpha c \varepsilon_0 \lambda_e} = \frac{\pi e^2}{\sqrt{\pi \varepsilon_0} \, \alpha c \, \lambda_e}$$

Energy equivalent for an electron:

$$c^2 m_e \quad = \qquad\qquad = \ 8.187,029,491,211,346,051,158,472,929,033,8\text{e-}14 \ J$$

$$= \frac{c^2 m_u \lambda_p}{\lambda_e} = \frac{\hbar R_\infty}{\alpha^2 c \varepsilon_0} = \frac{hc A_r(e)}{\lambda_p} = \frac{c^2 A_r(e)}{N_A} = \frac{c^2 A_r(e) K_B}{R} = \frac{hc \mu_0 R_\infty}{\alpha^2} = \frac{\sqrt{\varepsilon_0 \mu_0} \, 4\pi hc^2 R_\infty}{\alpha^2}$$

$$= \frac{\pi h A_r(e)}{\sqrt{\pi \varepsilon_0} \, \lambda_p} = \frac{\sqrt{\pi \varepsilon_0} \, e^2 c A_r(e)}{\alpha \varepsilon_0 \lambda_p} = \frac{\pi e^2 c A_r(e)}{\sqrt{\pi \varepsilon_0} \, \alpha \lambda_p} = \ = \frac{\pi h}{\sqrt{\pi \varepsilon_0} \, \lambda_e} = \frac{\sqrt{\pi \varepsilon_0} \, e^2 c}{\alpha \varepsilon_0 \lambda_e} = \frac{\pi e^2 c}{\sqrt{\pi \varepsilon_0} \, \alpha \, \lambda_e}$$

$$= \frac{\sqrt{\pi \varepsilon_0} \, h}{\varepsilon_0 \lambda_e} = \frac{\sqrt{\pi \varepsilon_0} \, h A_r(e)}{\varepsilon_0 \lambda_p}$$

Compton angular frequency (electron):

$$\omega_e \qquad = \qquad 1.552,673,809,564,159,342,339,165,278,999,1\text{e}21 \ radians/sec$$

$$\omega_e \quad = \quad \frac{2 m_e c^2}{h} \quad = \quad \frac{2 m_e}{h \varepsilon_0 \mu_0}$$

Compton wavelength of an electron:

$$\lambda_e = \frac{h}{c\, m_e} = 2.426,332,634,331,740,911,977,561,834,895,1\text{e-}12 \text{ m}$$

$$= \frac{m_u \lambda_p}{m_e} = \frac{\lambda_p}{m_e N_A} = \frac{\lambda_p}{A_r(e)} = \frac{K_B \lambda_p}{m_e R} = \frac{A_r(e)\lambda_e}{m_e N_A} = \frac{\pi h}{\sqrt{\pi \varepsilon_0}\, c^2 m_e} = \frac{\sqrt{\pi \varepsilon_0}\, e^2}{\alpha c \varepsilon_0 m_e}$$

$$= \frac{\pi e^2}{\sqrt{\pi \varepsilon_0}\, \alpha c\, m_e} = \frac{\sqrt{\pi \varepsilon_0}\, \hbar}{c^2 \varepsilon_0 m_e} = \frac{\sqrt{\pi \varepsilon_0}\, e^4}{\alpha^2 \hbar \varepsilon_0 m_e} = \frac{\sqrt{\pi \varepsilon_0}\, \sqrt{8c}\, \sqrt{2c}\, \pi e^2}{\alpha\, m_e} = \frac{\sqrt{\pi \varepsilon_0}\, \sqrt{8c}\, \sqrt{2c}\, \pi e^2}{\alpha c \sqrt{\varepsilon_0 \mu_0}\, m_e}$$

$$= \frac{\sqrt{\pi \varepsilon_0}\, \sqrt{8c}\, \sqrt{2c}\, \pi e^2}{\alpha\, c^2\, \varepsilon_0\, \mu_0\, m_e} = \frac{e^2}{\sqrt{\pi \varepsilon_0}\, \sqrt{8c}\, \sqrt{2c}\, \alpha\, c^2\, \varepsilon_0\, m_e} = \frac{\sqrt{\pi \varepsilon_0}\, \sqrt{8c}\, \sqrt{2c}\, \pi h}{c\, m_e}$$

Compton frequency (electron):

$$f_e = \frac{c^2 m_e}{h} = \frac{m_e}{h \varepsilon_0 \mu_0} = 1.235,578,559,007,300,580,211,788,226,959,8\text{e}20 \text{ sec}^{-1}$$

$$= \frac{\lambda_e}{c} = \frac{c\, m_e}{m_u \lambda_p} = \frac{c\, N_A m_e}{\lambda_p} = \frac{c\, A_r(e)}{\lambda_p} = \frac{c\, R\, m_e}{K_B \lambda_p} = \frac{c\, N_A m_e}{A_r(e)\lambda_e} = \frac{\sqrt{\pi \varepsilon_0}\, c\, m_e}{\pi h} = \frac{\alpha c^2 \varepsilon_0\, m_e}{\sqrt{\pi \varepsilon_0}\, e^2}$$

$$= \frac{\sqrt{\pi \varepsilon_0}\, \alpha c^2\, m_e}{\pi e^2} = \frac{c^3 \varepsilon_0\, m_e}{\sqrt{\pi \varepsilon_0}\, h} = \frac{\alpha^2 h c \varepsilon_0\, m_e}{\sqrt{\pi \varepsilon_0}\, e^4} = \frac{\alpha c^2 \sqrt{\varepsilon_0 \mu_0}\, m_e}{\sqrt{\pi \varepsilon_0}\, \sqrt{8c}\, \sqrt{2c}\, \pi e^2} = \frac{\alpha c^3 \varepsilon_0\, \mu_0\, m_e}{\sqrt{\pi \varepsilon_0}\, \sqrt{8c}\, \sqrt{2c}\, \pi e^2}$$

$$= \frac{\sqrt{\pi \varepsilon_0}\, \sqrt{8c}\, \sqrt{2c}\, \alpha c^3 \varepsilon_0\, m_e}{e^2} = \frac{c^2\, m_e}{\sqrt{\pi \varepsilon_0}\, \sqrt{8c}\, \sqrt{2c}\, \pi h} = \frac{\alpha c\, m_e}{\sqrt{\pi \varepsilon_0}\, \sqrt{8c}\, \sqrt{2c}\, \pi e^2}$$

Classical electron (orbit) radius:

$$r_e = 2.817,966,336,930,879,487,268,797,525,835,7\text{e-}15 \text{m}$$

$$r_e = \frac{e^2}{m_e} = \frac{\alpha\, \lambda_e}{2\pi} = \frac{e^2}{h f_e \varepsilon_0 \mu_0}$$

Bohr radius (electron orbit):

$$a_0 = \frac{\lambda_e}{2\pi\alpha} = \frac{2\lambda_e h \varepsilon_0}{e^2\sqrt{\varepsilon_0\mu_0}} = 5.291,820,953,494,772,831,141,276,511,983,5\text{e-}11 \text{ m}$$

$$= \frac{\hbar}{\alpha c\, m_e} = \frac{m_u \lambda_p}{\sqrt{\pi\varepsilon_0}\,\sqrt{8c}\,\sqrt{2c}\,\pi\alpha\, m_e} = \frac{\lambda_p}{\sqrt{\pi\varepsilon_0}\,\sqrt{8c}\,\sqrt{2c}\,\pi\alpha\, m_e N_A} = \frac{\lambda_p}{\sqrt{\pi\varepsilon_0}\,\sqrt{8c}\,\sqrt{2c}\,\pi\alpha\, A_r(e)}$$

$$= \frac{e^2}{\alpha^2 m_e} = \frac{K_B \lambda_p}{\sqrt{\pi\varepsilon_0}\,\sqrt{8c}\,\sqrt{2c}\,\pi\alpha\, m_e R} = \frac{A_r(e)\lambda_e}{\sqrt{\pi\varepsilon_0}\,\sqrt{8c}\,\sqrt{2c}\,\pi\alpha\, m_e N_A} = \frac{\hbar}{\sqrt{8c}\,\sqrt{2c}\,\pi\alpha c^2\varepsilon_0 m_e}$$

$$= \frac{e^2}{\alpha^2 c^2 \varepsilon_0 \mu_0 m_e} = \frac{e^2}{\sqrt{8c}\,\sqrt{2c}\,\pi\alpha^2 c\varepsilon_0 m_e} = \frac{\hbar}{\sqrt{8c}\,\sqrt{2c}\,\pi\alpha c^2\varepsilon_0 m_e} = \frac{e^4}{\sqrt{8c}\,\sqrt{2c}\,\pi\alpha^3 h\varepsilon_0 m_e}$$

$$= \frac{e^2}{\alpha^2 c\sqrt{\varepsilon_0\mu_0}\, m_e}$$

Relative mass of an electron:

$$A_r(e) = m_e N_A = \frac{\lambda_p}{\lambda_e} = 5.485,783,356,510,424,871,293,272,568,140,2\text{e-}4 \text{ Kg}$$

$$A_r(e) = \frac{m_e}{m_u} = \frac{\alpha^2 c\, m_e}{\alpha^2 c\, m_u} = \frac{\sqrt{8\varepsilon_0}\,\alpha^2 h\, \lambda_p m_e}{\sqrt{8\pi}\, e^4} = \frac{\sqrt{8\pi}\,\alpha^2 h\varepsilon_0\, \lambda_p m_e}{\sqrt{8\varepsilon_0}\,\pi e^4}$$

$$= \frac{R m_e}{K_B} = \frac{m_u A_r(e) R}{K_B} = \frac{e^2 R}{r_e K_B} = \frac{\hbar R R_\infty}{\alpha^2 c^3 \varepsilon_0 K_B} = \frac{e^2 c^2 \varepsilon_0 \lambda_p}{\sqrt{\pi\varepsilon_0}\,\hbar r_e} = \frac{\sqrt{\pi\varepsilon_0}\,e^2 c^2 \lambda_p}{\pi\hbar r_e} = \frac{\alpha c\varepsilon_0 \lambda_p}{\sqrt{\pi\varepsilon_0}\, r_e}$$

$$= \frac{\sqrt{\pi\varepsilon_0}\,\alpha c\, \lambda_p}{\pi r_e} = \frac{\sqrt{\pi\varepsilon_0}\,e^8 c^3 \lambda_p R_\infty}{\pi\alpha^6 h^4\varepsilon_0} = \frac{\alpha c\varepsilon_0 \lambda_p m_e}{\sqrt{\pi\varepsilon_0}\, e^2} = \frac{\alpha^2 h\varepsilon_0 \lambda_p m_e}{\sqrt{\pi\varepsilon_0}\, e^4} = \frac{\sqrt{\pi\varepsilon_0}\,\alpha c\, \lambda_p m_e}{\pi e^2}$$

$$= \frac{\sqrt{\pi\varepsilon_0}\,c^2 \lambda_p m_e}{\pi\hbar} = \frac{c^2\varepsilon_0 \lambda_p m_e}{\sqrt{\pi\varepsilon_0}\,\hbar} = \frac{\sqrt{8\varepsilon_0}\,c^2 \lambda_p m_e}{\sqrt{8\pi}\, h} = \frac{\sqrt{8\pi}\,c^2\varepsilon_0 \lambda_p m_e}{\sqrt{8\varepsilon_0}\,\pi h} = \frac{\sqrt{8\varepsilon_0}\,\alpha c\, \lambda_p m_e}{\sqrt{8\pi}\, e^2}$$

Electron frequency:

$$\alpha^2 c\, m_e \;=\; \frac{e^2 \alpha^2 c}{r_e} \;=\; \frac{\alpha^2 h}{\lambda_e} \;=\; 1.454{,}240{,}753{,}339{,}556{,}843{,}551{,}646{,}122{,}392{,}2e20 \text{ Hz}$$

$$\alpha^2 c\, m_e \;=\; \alpha^2 c\, m_u A_r(e) = 2h\mu_0 R_\infty = 4\pi\hbar R_\infty = \hbar R_\infty = \sqrt{\varepsilon_0 \mu_0}\; 4\pi\hbar c\, R_\infty = \frac{\hbar R_\infty}{c^2 \varepsilon_0} = \frac{\alpha^2 c\, A_r(e)}{N_A}$$

$$= \frac{\alpha^2 h\, A_r(e)}{\lambda_p} \;=\; \frac{\alpha^2 c\, A_r(e) K_B}{R} \;=\; \frac{\alpha^2 c\, m_u \lambda_p}{\lambda_e} \;=\; \frac{2\pi e^2 \alpha c}{\lambda_e} \;=\; \frac{\sqrt{\pi\varepsilon_0}\; \alpha^2 h\, A_r(e)}{c \varepsilon_0 \lambda_p} \;=\; \frac{\sqrt{\pi\varepsilon_0}\; \alpha^2 h}{c \varepsilon_0 \lambda_e}$$

$$= \frac{\pi\alpha^2 h\, A_r(e)}{\sqrt{\pi\varepsilon_0}\, c \lambda_p} \;=\; \frac{\sqrt{\pi\varepsilon_0}\; e^2 \alpha\, A_r(e)}{\varepsilon_0 \lambda_p} \;=\; \frac{\pi e^2 \alpha\, A_r(e)}{\sqrt{\pi\varepsilon_0}\, \lambda_p} \;=\; \frac{\pi\alpha^2 h}{\sqrt{\pi\varepsilon_0}\, c \lambda_e} \;=\; \frac{\sqrt{\pi\varepsilon_0}\; e^2 \alpha}{\varepsilon_0 \lambda_e} \;=\; \frac{\pi e^2 \alpha}{\sqrt{\pi\varepsilon_0}\, \lambda_e}$$

Mass of an electron:

$$m_e = m_u A_r(e) = \frac{e^2}{r_e} = \frac{h}{c \lambda_e} = 9.109{,}382{,}903{,}345{,}524{,}937{,}498{,}006{,}139{,}342{,}6e\text{-}31 \text{ Kg}$$

$$= \frac{m_u \lambda_p}{\lambda_e} = \frac{\hbar R_\infty}{\alpha^2 c^3 \varepsilon_0} = \frac{h\, A_r(e)}{c \lambda_p} = \frac{A_r(e)}{N_A} = \frac{A_r(e) K_B}{R} = \frac{\hbar \mu_0 R_\infty}{\alpha^2 c} = \frac{\sqrt{\varepsilon_0 \mu_0}\; 4\pi h\, R_\infty}{\alpha^2} = \frac{\sqrt{\pi\varepsilon_0}\; h\, A_r(e)}{c^2 \varepsilon_0 \lambda_p}$$

$$= \frac{\pi h\, A_r(e)}{\sqrt{\pi\varepsilon_0}\, c^2 \lambda_p} = \frac{\sqrt{\pi\varepsilon_0}\; e^2 A_r(e)}{\alpha c \varepsilon_0 \lambda_p} = \frac{\pi e^2 A_r(e)}{\sqrt{\pi\varepsilon_0}\, \alpha c \lambda_p} = \frac{\sqrt{\pi\varepsilon_0}\; h}{c^2 \varepsilon_0 \lambda_e} = \frac{\pi h}{\sqrt{\pi\varepsilon_0}\, c^2 \lambda_e} = \frac{\sqrt{\pi\varepsilon_0}\; e^2}{\alpha c \varepsilon_0 \lambda_e} = \frac{\pi e^2}{\sqrt{\pi\varepsilon_0}\, \alpha c \lambda_e}$$

Nucleon frequency:

$$\alpha^2 c\, m_u \;=\; \frac{\alpha^2 h}{\lambda_p} = \frac{\alpha^2 c}{N_A} \;=\; 2.650{,}926{,}328{,}714{,}177{,}930{,}460{,}751{,}597{,}717{,}3e23 \text{ Hz}$$

$$\alpha^2 c\, m_u \;=\; \frac{\alpha^2 c\, m_e}{A_r(e)} \;=\; \frac{\alpha^2 c\, K_B}{R} \;=\; \frac{2h R_\infty}{A_r(e)} \;=\; \frac{\alpha^2 c\, A_r(e) \lambda_e}{N_A \lambda_p} \;=\; \frac{\sqrt{\pi\varepsilon_0}\; e^4 c}{h \varepsilon_0 \lambda_p} \;=\; \frac{\sqrt{\pi\varepsilon_0}\; e^2 \alpha}{\varepsilon_0 \lambda_p}$$

$$= \frac{\alpha^2 c\, m_e \lambda_e}{\lambda_p} \;=\; \frac{\sqrt{\pi\varepsilon_0}\; \alpha^2 h}{c \varepsilon_0 \lambda_p} \;=\; \frac{\pi\alpha^2 h}{\sqrt{\pi\varepsilon_0}\, c \lambda_p} \;=\; \frac{4\pi h R_\infty}{A_r(e)} \;=\; \frac{\pi e^2 \alpha^2}{\sqrt{\pi\varepsilon_0}\, \alpha \lambda_p}$$

Frequency difference between an electron and a nucleon (proton/neutron):

$$\frac{\alpha^2 c\, m_u}{\alpha^2 c\, m_e} = \frac{1}{A_r(e)} = \frac{1}{m_e N_A} = 1,822.893,714,556,224,212,359,567,287,484,5 \text{ Hz}$$

$$= \frac{\lambda_e}{\lambda_p} = \frac{m_u}{m_e} = \frac{K_B}{R m_e} = \frac{r_e K_B}{e^2 R} = \frac{K_B}{m_u A_r(e) R} = \frac{\sqrt{8\pi}\, e^4}{\sqrt{8\varepsilon_0}\, \alpha^2 \hbar\, \lambda_p m_e} = \frac{\sqrt{8\varepsilon_0}\, \pi e^4}{\sqrt{8\pi}\, \alpha^2 \hbar \varepsilon_0\, \lambda_p m_e}$$

$$= \frac{\alpha^2 c^3 \varepsilon_0 K_B}{\hbar R R_\infty} = \frac{\sqrt{\pi\varepsilon_0}\, \hbar\, r_e}{e^2 c^2 \varepsilon_0\, \lambda_p} = \frac{\pi \hbar\, r_e}{\sqrt{\pi\varepsilon_0}\, e^2 c^2\, \lambda_p} = \frac{\sqrt{\pi\varepsilon_0}\, r_e}{\alpha c \varepsilon_0\, \lambda_p} = \frac{\pi\, r_e}{\sqrt{\pi\varepsilon_0}\, \alpha c\, \lambda_p} = \frac{\sqrt{\pi\varepsilon_0}\, e^2}{\alpha c \varepsilon_0\, \lambda_p m_e}$$

$$= \frac{\pi\alpha^6 \hbar^4 \varepsilon_0}{\sqrt{\pi\varepsilon_0}\, e^8 c^3\, \lambda_p R_\infty} = \frac{\sqrt{\pi\varepsilon_0}\, e^4}{\alpha^2 \hbar \varepsilon_0\, \lambda_p m_e} = \frac{\pi e^2}{\sqrt{\pi\varepsilon_0}\, \alpha c\, \lambda_p m_e} = \frac{\pi \hbar}{\sqrt{\pi\varepsilon_0}\, c^2\, \lambda_p m_e} = \frac{\sqrt{\pi\varepsilon_0}\, \hbar}{c^2 \varepsilon_0\, \lambda_p m_e}$$

$$= \frac{\sqrt{8\pi}\, \hbar}{\sqrt{8\varepsilon_0}\, c^2\, \lambda_p m_e} = \frac{\sqrt{8\varepsilon_0}\, \pi\hbar}{\sqrt{8\pi}\, c^2 \varepsilon_0\, \lambda_p m_e} = \frac{\sqrt{8\varepsilon_0}\, \alpha c\, \lambda_p m_e}{\sqrt{8\pi}\, e^2}$$

Atomic mass unit (mass of a nucleon):

$$m_u = \frac{K_B}{R} = \frac{1}{N_A} = \frac{m_e}{A_r(e)} = 1.660,528,356,720,136,653,603,011,597,202,5e\text{-}27 \text{ Kg}$$

$$m_u = \frac{h}{c\, \lambda_p} = \frac{A_r(e)\, \lambda_e}{N_A\, \lambda_p} = \frac{m_e \lambda_e}{\lambda_p} = \frac{\sqrt{\pi\varepsilon_0}\, \hbar}{c^2 \varepsilon_0\, \lambda_p} = \frac{\pi\hbar}{\sqrt{\pi\varepsilon_0}\, c^2\, \lambda_p} = \frac{\sqrt{\pi\varepsilon_0}\, e^2}{\alpha c \varepsilon_0\, \lambda_p} = \frac{\sqrt{\pi\varepsilon_0}\, e^4}{\alpha^2 \hbar \varepsilon_0\, \lambda_p}$$

$$m_u = \frac{\pi e^2}{\sqrt{\pi\varepsilon_0}\, \alpha c\, \lambda_p} = \frac{4\pi \hbar R_\infty}{A_r(e)\, \alpha^2 c} = \frac{\sqrt{\pi\varepsilon_0}\, \hbar}{c^2 \varepsilon_0\, \lambda_p}$$

Energy equivalent for a nucleon:

$$c^2\, m_u = \qquad = 1.492,408,460,041,560,499,244,908,258,965,1e\text{-}10 \text{ J}$$

$$= \frac{c^2 K_B}{R} = \frac{c^2}{N_A} = \frac{c^2 m_e}{A_r(e)} = \frac{4\pi \hbar c\, R_\infty}{A_r(e)\alpha^2} = \frac{hc}{\lambda_p} = \frac{c^2 m_e \lambda_e}{\lambda_p} = \frac{\sqrt{\pi\varepsilon_0}\, \hbar}{\varepsilon_0\, \lambda_p} = \frac{\pi\hbar}{\sqrt{\pi\varepsilon_0}\, \lambda_p}$$

$$= \frac{\sqrt{\pi\varepsilon_0}\, e^2 c}{\alpha \varepsilon_0\, \lambda_p} = \frac{\sqrt{\pi\varepsilon_0}\, e^4 c^2}{\alpha^2 \hbar \varepsilon_0\, \lambda_p} = \frac{\pi e^2 c}{\sqrt{\pi\varepsilon_0}\, \alpha\, \lambda_p}$$

Compton wavelength (nucleon):

$$\lambda_p = \frac{h}{m_u c} = A_r(e)\lambda_e = 1.331{,}025{,}054{,}892{,}506{,}909{,}053{,}994{,}151{,}094{,}2e\text{-}15 \text{ m}$$

$$\lambda_p = \frac{m_e \lambda_e}{m_u} = \frac{\sqrt{\pi\varepsilon_0}\, h}{c^2 \varepsilon_0 m_u} = \frac{\pi h}{\sqrt{\pi\varepsilon_0}\, c^2 m_u} = \frac{\sqrt{\pi\varepsilon_0}\, e^2}{\alpha c \varepsilon_0 m_u} = \frac{\pi e^2}{\sqrt{\pi\varepsilon_0}\,\alpha c m_u} = \frac{\sqrt{\pi\varepsilon_0}\, e^4}{\alpha^2 \hbar \varepsilon_0\, m_u}$$

$$\lambda_p = \frac{\sqrt{\pi\varepsilon_0}\, e^2 N_A}{\alpha c \varepsilon_0} = \frac{\pi e^2 N_A}{\sqrt{\pi\varepsilon_0}\,\alpha c} = \frac{\sqrt{\pi\varepsilon_0}\, e^4 N_A}{\alpha^2 \hbar \varepsilon_0} = \frac{\sqrt{8\pi}\, \hbar}{\sqrt{8\varepsilon_0}\, c^2\, m_u} = \frac{\sqrt{8\varepsilon_0}\,\pi h}{\sqrt{8\pi}\, c^2 \varepsilon_0 m_u} = \frac{\sqrt{\pi\varepsilon_0}\,\hbar N_A}{c^2 \varepsilon_0}$$

$$\lambda_p = \frac{\sqrt{8\pi}\, e^2}{\sqrt{8\varepsilon_0}\,\alpha c\, m_u} = \frac{\sqrt{8\pi}\, e^4}{\sqrt{8\varepsilon_0}\,\alpha^2 \hbar\, m_u} = \frac{\sqrt{8\varepsilon_0}\,\pi e^4}{\sqrt{8\pi}\,\alpha^2 \hbar \varepsilon_0 m_u} = \frac{\sqrt{8\pi}\,\hbar N_A}{\sqrt{8\varepsilon_0}\, c^2} = \frac{\sqrt{8\varepsilon_0}\,\pi h N_A}{\sqrt{8\pi}\, c^2 \varepsilon_0} = \frac{\pi h N_A}{\sqrt{\pi\varepsilon_0}\, c^2}$$

Rydberg constant:

$$R_\infty = \frac{\alpha^2}{2\lambda_e} = \frac{c\alpha^2 m_e}{\hbar\mu_0} = {}^*1.097{,}373{,}156{,}866{,}196{,}496{,}466{,}432{,}160{,}158{,}9$$

$$= \frac{\sqrt{\pi\varepsilon_0}\,\alpha^2 c\, A_r(e)}{\lambda_p} = \frac{\pi\alpha^2 c\varepsilon_0\, A_r(e)}{\sqrt{\pi\varepsilon_0}\,\lambda_p} = \frac{\sqrt{\pi\varepsilon_0}\, e^2 \alpha c^2\, A_r(e)}{\hbar\,\lambda_p} = \frac{\pi e^2 \alpha c^2 \varepsilon_0 A_r(e)}{\sqrt{\pi\varepsilon_0}\,\hbar\,\lambda_p} = \frac{\sqrt{\pi\varepsilon_0}\,\alpha^2 c}{\lambda_e}$$

$$= \frac{\sqrt{\pi\varepsilon_0}\, e^2 \alpha c^2}{\hbar\,\lambda_e} = \frac{\pi e^2 \alpha c^2 \varepsilon_0}{\sqrt{\pi\varepsilon_0}\,\hbar\,\lambda_e} = \frac{\alpha^2 c^3 \varepsilon_0\, m_u\, A_r(e)}{\hbar} = \frac{\alpha^2 c^3 \varepsilon_0\, A_r(e) K_B}{\hbar R} = \frac{e^2 \alpha^2 c^3 \varepsilon_0}{\hbar\, r_e} = \frac{\alpha^2 c^3 \varepsilon_0 A_r(e)}{\hbar N_A}$$

$$= \frac{m_e \alpha^2 c^3 \varepsilon_0}{\hbar} = \frac{\pi\alpha^2 c\varepsilon_0}{\sqrt{\pi\varepsilon_0}\,\lambda_e} = \frac{\alpha^2 m_e}{4\pi\hbar\sqrt{\varepsilon_0\mu_0}} = \frac{e^4 m_e}{8h^3 c\varepsilon_0^2}$$

Gas Constant: $R = 8.314{,}462{,}164{,}663{,}558{,}038{,}800{,}395{,}847{,}263{,}6 \text{ J/Mol} \cdot \text{K}$

$$R = K_B N_A = \frac{c\,\lambda_p K_B}{h} = \frac{K_B}{m_u} = \frac{A_r(e)\, K_B}{m_e} = \frac{\lambda_p K_B}{m_e \lambda_e} = \frac{c^2 \varepsilon_0 \lambda_p K_B}{\sqrt{\pi\varepsilon_0}\, h} = \frac{\sqrt{\pi\varepsilon_0}\, c^2 \lambda_p K_B}{\pi h}$$

$$= \frac{\alpha c\varepsilon_0 \lambda_p K_B}{\sqrt{\pi\varepsilon_0}\, e^2} = \frac{\sqrt{\pi\varepsilon_0}\,\alpha c\,\lambda_p K_B}{\pi e^2} = \frac{\alpha^2 \hbar \varepsilon_0 \lambda_p K_B}{\sqrt{\pi\varepsilon_0}\, e^4}$$

Avogadro number:

$$N_A = \frac{c\,\lambda_p}{h} = \frac{R}{K_B} = \frac{1}{m_u} = \frac{A_r(e)}{m_e} = 6.022,141,310,907,101,425,969,05 \; \text{Mol}^{-1} =$$

$$\frac{\lambda_p}{m_e\lambda_e} = \frac{c^2\varepsilon_0\,\lambda_p}{\sqrt{\pi\varepsilon_0}\,h} = \frac{\sqrt{\pi\varepsilon_0}\,c^2\lambda_p}{\pi\hbar} = \frac{\alpha c\varepsilon_0\,\lambda_p}{\sqrt{\pi\varepsilon_0}\,e^2} = \frac{\sqrt{\pi\varepsilon_0}\,\alpha c\,\lambda_p}{\pi e^2} = \frac{\alpha^2\hbar\varepsilon_0\,\lambda_p}{\sqrt{\pi\varepsilon_0}\,e^4}$$

QUANTUM JUMP NUMBER

Under certain circumstances, quantum constants and/or the math they produce can differ by certain arbitrary value which is listed below. This value first appeared in the mathematics regarding these physical constants and appears quite frequently in those particular mathematics and relationships. At first it was suspected that perhaps it was a quirk in the computer or the calculator which was being used, but different computers and calculators were tried with the same result.

For instance, the product of $\alpha \hbar$ divided by the product of $c\,e^2$ can yield this particular value which is thought to represent the initial quantum jump in the primordial singularity that produced the existence of our universe and the birth of four-dimensional space-time. This number has been named a "quantum jump number" and it appears quite frequently in the mathematics associated with the physical constants related to atoms and electromagnetic waves.

When you divide the numerical values which are represented by the numerator and denominator in these formulas directly, you get "1" or unity. However, as these formulas are derived down from higher numbers and formulas, you may get the following numerical value for each of these particular formulas. This process may quite possibly provide us with a detailed analysis of how the physical expansion process of the primordial singularity actually occurred, and how all other quantum changes-of-state and dual existences are able to take place.

Here are some examples:

$$\frac{e^2c}{\alpha\hbar} = \frac{\alpha\hbar}{e^2c} = \frac{e^4c^2}{\alpha^2\hbar^2} = \frac{\alpha^2\hbar^2}{e^4c^2} = .000,000,000,000,006,537,188,612,504,934,7$$

$$= \frac{\alpha^2\hbar^6}{e^{14}c^7} = \frac{e^{14}c^7}{\alpha^2\hbar^6} = \frac{\alpha^3\hbar^3}{e^6c^3} = \frac{e^6c^3}{\alpha^3\hbar^3} = \frac{\alpha^3\hbar^5}{e^6c} = \frac{e^6c}{\alpha^3\hbar^5} = \frac{\alpha^4\hbar^4}{e^8c^4} = \frac{e^8c^4}{\alpha^4\hbar^4}$$

$$= \frac{\alpha^5\hbar^5}{e^{10}c^5} = \frac{e^{10}c^5}{\alpha^5\hbar^5} = \frac{\alpha^5\hbar^6}{e^{10}c^8} = \frac{e^{10}c^8}{\alpha^5\hbar^6} = \frac{\alpha^5\hbar^7}{e^{10}c^3} = \frac{e^{10}c^3}{\alpha^5\hbar^7} = \frac{e^{12}c^6}{\alpha^6\hbar^6} = \frac{\alpha^6\hbar^6}{e^{12}c^6}$$

$$= \frac{\sqrt{2c}}{\sqrt{\pi\varepsilon_0}\sqrt{8c}\,c}\,\wedge= \frac{\sqrt{\pi\varepsilon_0}\sqrt{8c}\,c}{\sqrt{2c}}\,\wedge=\wedge\frac{\sqrt{8\pi}\sqrt{8c}\,c\varepsilon_0}{\sqrt{8\varepsilon_0}\sqrt{2c}} = \frac{\sqrt{8\varepsilon_0}\sqrt{8c}\,\pi c}{\sqrt{8\pi}\sqrt{2c}}\,\wedge= \frac{\pi c\varepsilon_0}{(\sqrt{\pi\varepsilon_0}-\pi c\varepsilon_0)}$$

$$= \frac{\alpha^4\hbar^4}{(\pi c\varepsilon_0 - \sqrt{\pi\varepsilon_0})\sqrt{\pi\varepsilon_0}\,e^8c^6} = \frac{(\pi c^3\varepsilon_0-c)}{(\sqrt{\pi\varepsilon_0}+\pi c\varepsilon_0)\sqrt{\pi\varepsilon_0}\,c^3} = \frac{\sqrt{8c}\sqrt{2c}\,\pi\alpha^4\hbar^4\varepsilon_0}{(\pi c\varepsilon_0-\sqrt{\pi\varepsilon_0})\sqrt{\pi\varepsilon_0}e^8c^5}$$

$$= \frac{\sqrt{8\pi}\,\alpha^4\hbar^4}{(\pi c\varepsilon_0 - \sqrt{\pi\varepsilon_0})\sqrt{8\varepsilon_0}\,\pi e^8c^6} = \frac{\sqrt{8\varepsilon_0}\,\alpha^4\hbar^4}{(\pi c\varepsilon_0 - \sqrt{\pi\varepsilon_0})\sqrt{8\pi}\,e^8c^6\varepsilon_0} = \frac{\sqrt{8\pi}\,e^6}{\sqrt{8\varepsilon_0}\sqrt{8c}\sqrt{2c}\,\pi\alpha\hbar^2c\varepsilon_0\,\lambda_p m_u}$$

$$= \frac{\sqrt{8\varepsilon_0}}{(\pi c\varepsilon_0 - \sqrt{\pi\varepsilon_0})\sqrt{8\pi}\,c^2\varepsilon_0} = \frac{\sqrt{8\pi}}{(\pi c\varepsilon_0 - \sqrt{\pi\varepsilon_0})\sqrt{8\varepsilon_0}\,\pi c^2} = \frac{\sqrt{8\pi}\sqrt{8c}\,\alpha^4\hbar^3\varepsilon_0^2\,\lambda_p m_u}{\sqrt{\pi\varepsilon_0}\sqrt{8\varepsilon_0}\sqrt{2c}\,e^8c}$$

$$= \frac{\sqrt{8c}\,\pi e^{20}c^{11}\varepsilon_0}{\sqrt{\pi\varepsilon_0}\sqrt{2c}\,\alpha^{10}\hbar^{10}} = \frac{\sqrt{8c}\,\pi\alpha^4\hbar^4\varepsilon_0}{\sqrt{\pi\varepsilon_0}\sqrt{2c}\,e^8c^3} = \frac{\sqrt{\pi\varepsilon_0}\sqrt{8c}\,e^{20}c^{11}}{\sqrt{2c}\,\alpha^{10}\hbar^{10}} = \frac{\sqrt{\pi\varepsilon_0}\sqrt{8c}\,\alpha^4\hbar^4}{\sqrt{2c}\,e^8c^3}$$

$$= \frac{\pi\hbar}{\sqrt{\pi\varepsilon_0}\,c^2\,\lambda_p m_u}\,\wedge= \frac{\pi e^2}{\sqrt{\pi\varepsilon_0}\,\alpha c\,\lambda_p m_u} = \frac{\sqrt{8\varepsilon_0}\,\pi}{\sqrt{\pi\varepsilon_0}\sqrt{8\pi}} = \frac{\sqrt{\pi\varepsilon_0}\sqrt{8\pi}}{\sqrt{8\varepsilon_0}\,\pi} = \frac{\sqrt{8\pi}\,\varepsilon_0}{\sqrt{\pi\varepsilon_0}\sqrt{8\varepsilon_0}} = \frac{\sqrt{\pi\varepsilon_0}\sqrt{8\varepsilon_0}}{\sqrt{8\pi}\,\varepsilon_0}$$

$$=\wedge\frac{\sqrt{8\varepsilon_0}\,c^2\,\lambda_p m_e}{\sqrt{8\pi}\,\hbar\,A_r(e)} =\wedge\frac{\sqrt{8\pi}\,e^2c^3\varepsilon_0\,\lambda_p m_u}{\sqrt{8\varepsilon_0}\,\pi\alpha\hbar^2} =\wedge\frac{\sqrt{8\pi}\,c^2\varepsilon_0\,\lambda_p m_u}{\sqrt{8\varepsilon_0}\,\pi\hbar} = \frac{\sqrt{8\pi}\,e^6}{\sqrt{8\varepsilon_0}\sqrt{8c}\sqrt{2c}\,\pi\alpha\hbar^2c\varepsilon_0\,\lambda_e m_e}$$

> This symbol represents a "continuance" or "continuation" to the numerical values within the brackets.

V This symbol indicates that a "quantum jump" down to the "quantum jump number" has taken place.

† This symbol indicates that a "quantum jump" all the way down to "unity" has taken place which is represented by the numerical values within the large brackets.

$\int \ddagger \int$ This symbol indicates that all the conditions have been met and the process has been initiated for a "quantum jump" or "jumps" to take place.

$\int \dagger \int$ This symbol indicates that a "quantum jump" which includes, but moves below the "quantum jump number" all the way down to "unity", is in the process of taking place.

$\int X \int$ This symbol indicates that a "quantum reversal" has taken place during the initiation process whereby the numerator and the denominator are reversed but the numerical value stays the same.

$\int\int\int$ These symbols represent the number of quantum jumps that are in the process of taking place.

$$\pi\alpha^6\hbar^8 \div \sqrt{\pi\varepsilon_0}\, e^8 c^2 \varepsilon_0 = \frac{\pi\alpha^6\hbar^8}{\sqrt{\pi\varepsilon_0}\, e^8 c^2 \varepsilon_0} > \int \ddagger \int > \begin{pmatrix} 7.256,894,923,231,724,801,589,587,189,309,4 \\ 1.259,297,666,035,085,735,991,190,862,792,4 \\ 1.259,297,666,035,077,503,724,828,704,241,4 \end{pmatrix}$$

$$\frac{\pi\alpha^6\hbar^8}{\sqrt{\pi\varepsilon_0}\, e^8 c^2 \varepsilon_0} \div \frac{\sqrt{\pi\varepsilon_0}\,\alpha^6\hbar^8}{e^8 c^2 \varepsilon_0^2} = \int \dagger \int > \int > \begin{cases} 1.259,297,666,035,085,735,991,190,862,792,4 \\ V\ 1.000,000,000,000,006,537,188,612,504,934,8 \\ \dagger\ 1.000,000,000,000,000,000,000,000,000,000,0 \\ 1.259,297,666,035,077,503,724,828,704,241,4 \end{cases}$$

$$\sqrt{\pi\varepsilon_0}\,\alpha^6\hbar^8 \div e^8 c^2 \varepsilon_0^2 = \frac{\sqrt{\pi\varepsilon_0}\,\alpha^6\hbar^8}{e^8 c^2 \varepsilon_0^2} > \int \ddagger \int > \begin{pmatrix} 3.852,567,560,358,045,371,299,054,355,542,9 \\ 1.259,297,666,035,077,503,724,828,704,241,4 \\ 3.059,298,579,094,430,572,855,128,522,780,7 \end{pmatrix}$$

$$\varepsilon_0^2 \lambda_p m_u \div \pi\alpha^2\hbar^5 = \frac{\varepsilon_0^2 \lambda_p m_u}{\pi\alpha^2\hbar^5} > \int \ddagger \int > \begin{pmatrix} 1.732,737,397,115,734,570,086,740,713,345,9 \\ 7.940,934,276,075,639,012,859,360,641,827,4 \\ 2.182,032,160,039,539,778,487,223,170,935,1 \end{pmatrix}$$

$$\frac{\varepsilon_0^2 \lambda_p m_u}{\pi\alpha^2\hbar^5} \div \frac{\sqrt{\pi\varepsilon_0}\, e^8 c^2 \varepsilon_0}{\pi\alpha^6\hbar^8} = \frac{\alpha^4\hbar^3 \varepsilon_0 \lambda_p m_u}{\sqrt{\pi\varepsilon_0}\, e^8 c^2} > \int\int > \begin{cases} 7.940,934,276,075,639,012,859,360,641,827,4 \\ 6.508,392,053,655,030,282,388,672,704,907,1 \\ V\ 1.000,000,000,000,006,537,188,612,504,934,9 \\ \dagger\ 1.000,000,000,000,000,000,000,000,000,000,0 \\ 6.508,392,053,655,030,282,388,672,704,907,1 \\ 7.940,934,276,075,587,101,474,238,430,381,6 \end{cases}$$

$$\sqrt{\pi\varepsilon_0}\, e^8 c^2 \varepsilon_0 \div \pi\alpha^6\hbar^8 = \frac{\sqrt{\pi\varepsilon_0}\, e^8 c^2 \varepsilon_0}{\pi\alpha^6\hbar^8} > \int \ddagger \int > \begin{pmatrix} 5.762,652,563,376,971,982,049,225,428,309,7 \\ 7.940,934,276,075,587,101,474,238,430,381,6 \\ 7.256,894,923,231,724,801,589,587,189,309,4 \end{pmatrix}$$

$$e^8\alpha^4\hbar^4 \div \sqrt{\pi\varepsilon_0}\,c^3\varepsilon_0 = \frac{e^8\alpha^4\hbar^4}{\sqrt{\pi\varepsilon_0}\,c^3\varepsilon_0} > \int \ddagger \int > \left\{\begin{array}{l} 3.728,843,520,996,486,004,389,516,040,711,3 \\ 9.371,605,929,558,766,692,941,114,074,518,7 \\ 3.978,873,577,297,383,394,222,094,084,312,1 \end{array}\right.$$

$$\frac{e^8\alpha^4\hbar^4}{\sqrt{\pi\varepsilon_0}\,c^3\varepsilon_0} \div \frac{e^{16}c}{\sqrt{\pi\varepsilon_0}\,\varepsilon_0} = \frac{\alpha^4\hbar^4}{e^8c^4} > \iiint > \left\{\begin{array}{l} 9.371,605,929,558,766,692,941,114,074,518,7 \\ 3.507,242,624,537,976,003,177,050,898,525,9 \\ \sqrt{2c} = 2.448,642,309,525,831,908,821,171,317,036,3 \\ \vee \quad 1.000,000,000,000,006,537,188,612,504,934,3 \\ \dagger \quad 1.000,000,000,000,000,000,000,000,000,000,0 \\ 3.507,242,624,537,976,003,177,050,898,525,9 \\ 3.827,266,192,820,761,231,567,626,267,746,5 \end{array}\right\}$$

$$e^{16}c \div \sqrt{\pi\varepsilon_0}\,\varepsilon_0 = \frac{e^{16}c}{\sqrt{\pi\varepsilon_0}\,\varepsilon_0} > \int \ddagger \int > \left\{\begin{array}{l} 5.651,798,902,036,413,815,076,646,291,150,7 \\ 3.827,266,192,820,761,231,567,626,267,746,5 \\ 1.476,719,574,049,522,928,714,998,399,277,8 \end{array}\right.$$

$$\sqrt{\pi\varepsilon_0}\,e^8\alpha^4\hbar^4 \div \pi c^3\varepsilon_0^2 = \frac{\sqrt{\pi\varepsilon_0}\,e^8\alpha^4\hbar^4}{\pi c^3\varepsilon_0^2} > \int \ddagger \int > \left\{\begin{array}{l} 6.219,041,576,083,421,692,332,093,358,918,5 \\ 9.371,605,929,558,766,692,941,114,074,518,7 \\ 6.636,046,823,595,180,960,526,522,125,371,3 \end{array}\right.$$

$$\frac{\sqrt{\pi\varepsilon_0}\,e^8\alpha^4\hbar^4}{\pi c^3\varepsilon_0^2} \div \frac{\sqrt{\pi\varepsilon_0}\,e^{16}c}{\pi\varepsilon_0^2} = \frac{\alpha^4\hbar^4}{e^8c^4} > \iiint > \left\{\begin{array}{l} 9.371,605,929,558,766,692,941,114,074,518,7 \\ 3.507,242,624,537,976,003,177,050,898,525,9 \\ \sqrt{2c} = 2.448,642,309,525,831,908,821,171,317,036,3 \\ \vee \quad 1.000,000,000,000,006,537,188,612,504,934,3 \\ \dagger \quad 1.000,000,000,000,000,000,000,000,000,000,0 \\ 3.507,242,624,537,976,003,177,050,898,525,9 \\ 3.827,266,192,820,761,231,567,626,267,746,5 \end{array}\right\}$$

$$\sqrt{\pi\varepsilon_0}\,e^{16}c \div \pi\varepsilon_0^2 = \frac{\sqrt{\pi\varepsilon_0}\,e^{16}c}{\pi\varepsilon_0^2} > \int \ddagger \int > \left\{\begin{array}{l} 9.426,185,934,998,427,837,495,241,943,596,1 \\ 3.827,266,192,820,761,231,567,626,267,746,5 \\ 2.462,903,142,896,148,055,724,267,752,055,6 \end{array}\right.$$

$$\varepsilon_0^2\lambda_p \div \pi\alpha^2\hbar^5 N_A = \frac{\varepsilon_0^2\lambda_p}{\pi\alpha^2\hbar^5 N_A} > \int \ddagger \int > \left\{\begin{array}{l} 1.043,478,946,012,430,856,926,803,442,622,2 \\ 7.940,934,276,075,639,012,859,360,641,827,4 \\ 1.314,050,601,270,196,821,729,834,881,897,4 \end{array}\right.$$

$$\frac{\varepsilon_0^2\lambda_p}{\pi\alpha^2\hbar^5 N_A} \div \frac{\sqrt{\pi\varepsilon_0}\,e^8c^2\varepsilon_0}{\pi\alpha^6\hbar^8} = \frac{\alpha^4\hbar^3\varepsilon_0\lambda_p}{\sqrt{\pi\varepsilon_0}\,e^4c^2 N_A} > \iint > \left\{\begin{array}{l} 7.940,934,276,075,639,012,859,360,641,827,4 \\ 6.508,392,053,655,030,282,388,672,704,907,1 \\ \vee \quad 1.000,000,000,000,006,537,188,612,504,934,9 \\ \dagger \quad 1.000,000,000,000,000,000,000,000,000,000,0 \\ 6.508,392,053,655,030,282,388,672,704,907,1 \\ 7.940,934,276,075,587,101,474,238,430,381,6 \end{array}\right\}$$

$$\sqrt{\pi\varepsilon_0}\,e^8c^2\varepsilon_0 \div \pi\alpha^6\hbar^8 = \frac{\sqrt{\pi\varepsilon_0}\,e^8c^2\varepsilon_0}{\pi\alpha^6\hbar^8} > \int \ddagger \int > \left\{\begin{array}{l} 5.762,652,563,376,971,982,049,225,428,309,7 \\ 7.940,934,276,075,587,101,474,238,430,381,6 \\ 7.256,894,923,231,724,801,589,587,189,309,4 \end{array}\right.$$

$$\alpha\varepsilon_0 \;\div\; \sqrt{\pi\varepsilon_0}\,\pi e^8\hbar^3 c^5 \;=\; \frac{\alpha\varepsilon_0}{\sqrt{\pi\varepsilon_0}\,\pi e^8\hbar^3 c^5} \;>\; \int\ddagger\int \;>\; \begin{cases} 6.461,213,022,440,564,360,054,124,509,616,9 \\ 1.000,000,006,661,624,754,502,248,052,661,4 \\ 6.461,212,979,398,388,032,382,599,128,666,2 \end{cases}$$

$$\frac{\alpha\varepsilon_0}{\sqrt{\pi\varepsilon_0}\,\pi e^8\hbar^3 c^5} \;\div\; \frac{\sqrt{\pi\varepsilon_0}\,e^2}{\pi^2\alpha^4\hbar^8} \;=\; \frac{\alpha^5\hbar^5}{e^{10}c^5} \;>\; \iint \;>\; \begin{cases} 1.000,000,006,661,624,754,502,248,052,661,4 \\ 2.699,027,212,360,063,487,776,520,170,569,3 \\ \vee\ \ 1.000,000,000,000,006,537,188,612,504,934,9 \\ \dagger\ \ 1.000,000,000,000,000,000,000,000,000,000,0 \\ 2.699,027,212,360,063,487,776,520,170,569,3 \\ 1.000,000,006,661,618,217,313,591,999,471,7 \end{cases}$$

$$\sqrt{\pi\varepsilon_0}\,e^2 \;\div\; \pi^2\alpha^4\hbar^8 \;=\; \frac{\sqrt{\pi\varepsilon_0}\,e^2}{\pi^2\alpha^4\hbar^8} \;>\; \int\!\times\!\int \;>\; \begin{cases} 4.281,244,699,343,082,511,768,861,252,562,3 \\ 1.000,000,006,661,618,217,313,591,999,471,8 \\ 4.281,244,670,823,065,019,837,198,856,087,8 \end{cases}$$

$$\pi^5 e^{22}\alpha^{14}\hbar^8 \;\div\; \sqrt{8c}\,\varepsilon_0^2\,N_A \;=\; \frac{\pi^5 e^{22}\alpha^{14}\hbar^8}{\sqrt{8c}\,\varepsilon_0^2\,N_A} \;>\; \int\ddagger\int \;>\; \begin{cases} 1.812,596,801,042,356,580,652,594,214,228,4 \\ 7.839,664,220,416,710,818,319,233,740,611,9 \\ 2.312,084,739,958,428,357,600,557,121,213,7 \end{cases}$$

$$\frac{\pi^5 e^{22}\alpha^{14}\hbar^8}{\sqrt{8c}\,\varepsilon_0^2\,N_A} \;\div\; \frac{\pi^5 e^{18}\alpha^{16}\hbar^{10}}{\sqrt{8c}\,c^2\varepsilon_0^2\,N_A} \;=\; \frac{e^4 c^2}{\alpha^2\hbar^2} \;>\; \iint \;>\; \begin{cases} 7.839,664,220,416,710,818,319,233,740,611,9 \\ 5.922,197,754,666,738,586,193,956,454,205,6 \\ \vee\ \ 1.000,000,000,000,006,537,188,612,504,934,9 \\ \dagger\ \ 1.000,000,000,000,000,000,000,000,000,000,0 \\ 5.922,197,754,666,738,586,193,956,454,205,6 \\ 7.839,664,220,416,659,568,955,566,170,447,1 \end{cases}$$

$$\pi^5 e^{18}\alpha^{16}\hbar^{10} \div \sqrt{8c}\,c^2\varepsilon_0^2 N_A \;=\; \frac{\pi^5 e^{18}\alpha^{16}\hbar^{10}}{\sqrt{8c}\,c^2\varepsilon_0^2 N_A} \;>\; \int\ddagger\int \;>\; \begin{cases} 1.629,080,761,898,596,421,791,266,850,597,7 \\ 7.839,664,220,416,659,568,955,566,170,447,1 \\ 2.077,998,133,716,005,812,711,719,248,931,8 \end{cases}$$

$$\sqrt{\pi\varepsilon_0}\,e^8\alpha^4\hbar^4 \;\div\; \pi c^2\varepsilon_0^2 \;=\; \frac{\sqrt{\pi\varepsilon_0}\,e^8\alpha^4\hbar^4}{\pi c^2\varepsilon_0^2} \;>\; \int\ddagger\int \;>\; \begin{cases} 6.219,041,576,083,421,692,332,093,358,918,5 \\ 2.809,536,777,029,797,522,325,347,837,658,4 \\ 2.213,546,954,405,097,462,634,140,757,927,7 \end{cases}$$

$$\frac{\sqrt{\pi\varepsilon_0}\,e^8\alpha^4\hbar^4}{\pi c^3\varepsilon_0^2} \;\div\; \frac{e^{16}c^2}{\sqrt{\pi\varepsilon_0}\,\varepsilon_0} \;=\; \frac{\alpha^4\hbar^4}{e^8 c^4} \;>\; \iiint \;>\; \begin{cases} 2.809,536,777,029,797,522,325,347,837,658,4 \\ 3.507,242,624,537,976,003,177,050,898,525,9 \\ \sqrt{2c} = 2.448,642,309,525,831,908,821,171,317,036,3 \\ \vee\ \ 1.000,000,000,000,006,537,188,612,504,934,3 \\ \dagger\ \ 1.000,000,000,000,000,000,000,000,000,000,0 \\ 3.507,242,624,537,976,003,177,050,898,525,9 \\ 1.147,385,539,366,037,963,042,765,872,033,6 \end{cases}$$

$$e^{16}c^2 \;\div\; \sqrt{\pi\varepsilon_0}\,\varepsilon_0 \;=\; \frac{e^{16}c^2}{\sqrt{\pi\varepsilon_0}\,\varepsilon_0} \;>\; \int\ddagger\int \;>\; \begin{cases} 1.694,366,684,963,197,703,126,985,250,021,4 \\ 1.147,385,539,366,037,963,042,765,872,033,6 \\ 1.476,719,574,049,522,928,714,998,399,277,8 \end{cases}$$

$$\frac{\alpha^3 \varepsilon_0 \lambda_p}{e^8 \hbar} \div \pi^2\alpha^2\hbar^3 c^3 N_A = \frac{\alpha\varepsilon_0 \lambda_p}{\pi^2 e^8\hbar^4 c^3 N_A} > \int \ddagger \int > \left\{\begin{array}{l} 1.000,165,402,709,875,094,497,946,752,503,2 \\ \quad 1.000,000,006,661,624,754,502,248,052,661,4 \\ 1.000,165,396,047,148,533,593,717,207,496,4 \end{array}\right.$$

$$\frac{\alpha\varepsilon_0 \lambda_p}{\pi^2 e^8\hbar^4 c^3 N_A} \div \frac{\sqrt{\pi\varepsilon_0}\, e^2}{\pi^2\alpha^4\hbar^8} = \frac{\alpha^5\hbar^4\varepsilon_0\lambda_p}{\sqrt{\pi\varepsilon_0}\, e^{10}c^3 N_A} > \iint > \left\{\begin{array}{l} 1.000,000,006,661,624,754,502,248,052,661,1 \\ 3.016,241,429,731,898,878,424,985,668,485,3 \\ \text{V} \quad 1.000,000,000,000,006,537,188,612,504,934,9 \\ \dagger \quad 1.000,000,000,000,000,000,000,000,000,000,0 \\ 3.016,241,429,731,898,878,424,985,668,485,3 \\ 1.000,000,006,661,618,217,313,591,999,471,7 \end{array}\right\}$$

$$\sqrt{\pi\varepsilon_0}\, e^2 \div \pi^2\alpha^4\hbar^8 = \frac{\sqrt{\pi\varepsilon_0}\, e^2}{\pi^2\alpha^4\hbar^8} > \int X \int > \left\{\begin{array}{l} 4.281,244,699,343,082,511,768,861,252,562,3 \\ \quad 1.000,000,006,661,618,217,313,591,999,471,8 \\ 4.281,244,670,823,065,019,837,198,856,087,8 \end{array}\right.$$

$$\varepsilon_0^2 A_r(e)\lambda_e \div \pi\alpha^2\hbar^5 N_A = \frac{\varepsilon_0^2 A_r(e)\lambda_e}{\pi\alpha^2\hbar^5 N_A} > \int \ddagger \int > \left\{\begin{array}{l} 1.043,478,946,012,430,856,926,803,442,622,2 \\ \quad 7.940,934,276,075,639,012,859,360,641,827,4 \\ 1.314,050,601,270,196,821,729,834,881,897,4 \end{array}\right.$$

$$\frac{\varepsilon_0^2 A_r(e)\lambda_e}{\pi\alpha^2\hbar^5 N_A} \div \frac{\sqrt{\pi\varepsilon_0}\, e^8 c^2\varepsilon_0}{\pi\alpha^6\hbar^8} = \frac{\alpha^4\hbar^3\varepsilon_0 A_r(e)\lambda_e}{\sqrt{\pi\varepsilon_0}\, e^4 c^2 N_A} > \iint > \left\{\begin{array}{l} 7.940,934,276,075,639,012,859,360,641,827,4 \\ 6.508,392,053,655,030,282,388,672,704,907,1 \\ \text{V} \quad 1.000,000,000,000,006,537,188,612,504,934,9 \\ \dagger \quad 1.000,000,000,000,000,000,000,000,000,000,0 \\ 6.508,392,053,655,030,282,388,672,704,907,1 \\ 7.940,934,276,075,587,101,474,238,430,381,6 \end{array}\right\}$$

$$\sqrt{\pi\varepsilon_0}\, e^8 c^2\varepsilon_0 \div \pi\alpha^6\hbar^8 = \frac{\sqrt{\pi\varepsilon_0}\, e^8 c^2\varepsilon_0}{\pi\alpha^6\hbar^8} > \int \ddagger \int > \left\{\begin{array}{l} 5.762,652,563,376,971,982,049,225,428,309,7 \\ \quad 7.940,934,276,075,587,101,474,238,430,381,6 \\ 7.256,894,923,231,724,801,589,587,189,309,4 \end{array}\right.$$

$$e^8\alpha^4\hbar^4 \div \sqrt{\pi\varepsilon_0}\, c^2\varepsilon_0 = \frac{e^8\alpha^4\hbar^4}{\sqrt{\pi\varepsilon_0}\, c^2\varepsilon_0} > \int \ddagger \int > \left\{\begin{array}{l} 3.728,843,520,996,486,004,389,516,040,711,3 \\ \quad 2.809,536,777,029,797,522,325,347,837,658,4 \\ 1.327,209,364,719,036,192,105,304,425,074,3 \end{array}\right.$$

$$\frac{e^8\alpha^4\hbar^4}{\sqrt{\pi\varepsilon_0}\, c^2\varepsilon_0} \div \frac{\sqrt{\pi\varepsilon_0}\, e^{16}c^2}{\pi\varepsilon_0^2} = \frac{\alpha^4\hbar^4}{e^8 c^4} > \iiint > \left\{\begin{array}{l} 2.809,536,777,029,797,522,325,347,837,658,4 \\ 3.507,242,624,537,976,003,177,050,898,525,9 \\ \sqrt{2c} = 2.448,642,309,525,831,908,821,171,317,036,3 \\ \text{V} \quad 1.000,000,000,000,006,537,188,612,504,934,3 \\ \dagger \quad 1.000,000,000,000,000,000,000,000,000,000,0 \\ 3.507,242,624,537,976,003,177,050,898,525,9 \\ 1.147,385,539,366,037,963,042,765,872,033,6 \end{array}\right\}$$

$$\sqrt{\pi\varepsilon_0}\, e^{16}c^2 \div \pi\varepsilon_0^2 = \frac{\sqrt{\pi\varepsilon_0}\, e^{16}c^2}{\pi\varepsilon_0^2} > \int \ddagger \int > \left\{\begin{array}{l} 2.825,899,451,018,206,907,538,323,145,576,6 \\ \quad 1.147,385,539,366,037,963,042,765,872,033,6 \\ 2.462,903,142,896,148,055,724,267,752,055,6 \end{array}\right.$$

$$\varepsilon_0^2 \lambda_e m_e \div \pi\alpha^2\hbar^5 = \frac{\varepsilon_0^2 \lambda_e m_e}{\pi\alpha^2\hbar^5} > \int \ddagger \int > \begin{Bmatrix} 1.732,737,397,115,734,570,086,740,713,345,9 \\ 7.940,934,276,075,639,012,859,360,641,827,4 \\ 2.182,032,160,039,539,778,487,223,170,935,1 \end{Bmatrix}$$

$$\frac{\varepsilon_0^2 \lambda_e m_e}{\pi\alpha^2\hbar^5} \div \frac{\sqrt{\pi\varepsilon_0}\, e^8 c^2\varepsilon_0}{\pi\alpha^6\hbar^8} = \frac{\alpha^4\hbar^3\varepsilon_0\, \lambda_e m_e}{\sqrt{\pi\varepsilon_0}\, e^4 c^2} > \iint > \begin{Bmatrix} 7.940,934,276,075,639,012,859,360,641,827,4 \\ 6.508,392,053,655,030,282,388,672,704,907,1 \\ \vee\ 1.000,000,000,000,006,537,188,612,504,934,9 \\ \dagger\ 1.000,000,000,000,000,000,000,000,000,000,0 \\ 6.508,392,053,655,030,282,388,672,704,907,1 \\ 7.940,934,276,075,587,101,474,238,430,381,6 \end{Bmatrix}$$

$$\sqrt{\pi\varepsilon_0}\, e^8 c^2\varepsilon_0 \div \pi\alpha^6\hbar^8 v = \frac{\sqrt{\pi\varepsilon_0}\, e^8 c^2\varepsilon_0}{\pi\alpha^6\hbar^8} > \int \ddagger \int > \begin{Bmatrix} 5.762,652,563,376,971,982,049,225,428,309,7 \\ 7.940,934,276,075,587,101,474,238,430,381,6 \\ 7.256,894,923,231,724,801,589,587,189,309,4 \end{Bmatrix}$$

$$\sqrt{\pi\varepsilon_0}\alpha^{12}\hbar^8 c^9\varepsilon_0^5 \div \pi^6 e^{18}\lambda_p = \frac{\sqrt{\pi\varepsilon_0}\alpha^{12}\hbar^8 c^9\varepsilon_0^5}{\pi^6 e^{18}\lambda_p} > \int \ddagger \int > \begin{Bmatrix} 6.192,588,443,292,445,130,375,825,985,603,7 \\ 1.000,000,008,433,980,347,412,782,131,980,3 \\ 6.192,588,391,064,276,340,523,179,077,314,1 \end{Bmatrix}$$

$$\frac{\sqrt{\pi\varepsilon_0}\,\alpha^{12}\hbar^8 c^9\varepsilon_0^5}{\pi^6 e^{18}\lambda_p} \div \frac{\sqrt{\pi\varepsilon_0}\,\alpha^3 c^{18}\varepsilon_0^5}{\pi^6\hbar\,\lambda_p} = \frac{\alpha^9\hbar^9}{e^{18}c^9} > \iint > \begin{Bmatrix} 1.000,000,008,433,980,347,412,782,131,980,3 \\ 9.466,143,283,977,126,171,847,478,873,690,6 \\ \vee\ 1.000,000,000,000,006,537,188,612,504,934,9 \\ \dagger\ 1.000,000,000,000,000,000,000,000,000,000,0 \\ 9.466,143,283,977,126,171,847,478,873,690,6 \\ 1.000,000,008,433,973,810,224,114,492,568,4 \end{Bmatrix}$$

$$\sqrt{\pi\varepsilon_0}\,\alpha^3 c^{18}\varepsilon_0^5 \div \pi^6\hbar\,\lambda_p = \frac{\sqrt{\pi\varepsilon_0}\,\alpha^3 c^{18}\varepsilon_0^5}{\pi^6\hbar\,\lambda_p} > \int \ddagger \int > \begin{Bmatrix} 1.349,464,901,828,688,175,636,534,192,738,6 \\ 1.000,000,008,433,973,810,224,114,492,568,4 \\ 1.349,464,890,447,336,631,786,743,216,834,2 \end{Bmatrix}$$

$$\varepsilon_0^2 K_B\lambda_p \div \pi\alpha^2\hbar^5 R = \frac{\varepsilon_0^2 K_B\lambda_p}{\pi\alpha^2\hbar^5 R} > \int \ddagger \int > \begin{Bmatrix} 1.440,677,952,961,638,964,103,923,136,649,9 \\ 7.940,934,276,075,639,012,859,360,641,827,9 \\ 1.814,242,383,672,785,121,292,018,670,949,1 \end{Bmatrix}$$

$$\frac{\varepsilon_0^2 K_B\lambda_p}{\pi\alpha^2\hbar^5 R} \div \frac{\sqrt{\pi\varepsilon_0}\, e^8 c^2\varepsilon_0}{\pi\alpha^6\hbar^8} = \frac{\alpha^4\hbar^3\varepsilon_0\, K_B\lambda_p}{\sqrt{\pi\varepsilon_0}\, e^8 c^2 R} > \iint > \begin{Bmatrix} 7.940,934,276,075,639,012,859,360,641,827,9 \\ 5.411,377,948,291,170,305,806,226,734,225,4 \\ \vee\ 1.000,000,000,000,006,537,188,612,504,934,9 \\ \dagger\ 1.000,000,000,000,000,000,000,000,000,000,0 \\ 5.411,377,948,291,170,305,806,226,734,225,4 \\ 7.940,934,276,075,587,101,474,238,430,381,6 \end{Bmatrix}$$

$$\sqrt{\pi\varepsilon_0}\, e^8 c^2\varepsilon_0 \div \pi\alpha^6\hbar^8 = \frac{\sqrt{\pi\varepsilon_0}\, e^8 c^2\varepsilon_0}{\pi\alpha^6\hbar^8} > \int \ddagger \int > \begin{Bmatrix} 5.762,652,563,376,971,982,049,225,428,309,7 \\ 7.940,934,276,075,587,101,474,238,430,381,6 \\ 7.256,894,923,231,724,801,589,587,189,309,4 \end{Bmatrix}$$

$$\alpha\varepsilon_0 \;\div\; \sqrt{\pi\varepsilon_0}\,\pi e^8\hbar^3 c^5 \;=\; \frac{\alpha\varepsilon_0}{\sqrt{\pi\varepsilon_0}\,\pi e^8\hbar^3 c^5} \;>\; \int\ddagger\int \;>\; \left\{\begin{array}{l} 6.461,213,022,440,564,360,054,124,509,616,9 \\ 1.000,000,006,661,624,754,502,248,052,661,4 \\ 6.461,212,979,398,388,032,382,599,128,666,2 \end{array}\right.$$

$$\frac{\alpha\varepsilon_0}{\sqrt{\pi\varepsilon_0}\,\pi e^8\hbar^3 c^5} \;\div\; \frac{e^2\varepsilon_0}{\sqrt{\pi\varepsilon_0}\,\pi\alpha^4\hbar^8} \;=\; \frac{\alpha^5\hbar^5}{e^{10}c^5} \;>\; \iint \;>\; \left\{\begin{array}{l} 1.000,000,006,661,624,754,502,248,052,661,4 \\ 2.699,027,212,360,063,487,776,520,170,569,3 \\ \vee\;\; 1.000,000,000,000,006,537,188,612,504,934,9 \\ \dagger\;\; 1.000,000,000,000,000,000,000,000,000,000,0 \\ 2.699,027,212,360,063,487,776,520,170,569,3 \\ 1.000,000,006,661,618,217,313,591,999,471,7 \end{array}\right.$$

$$e^2\varepsilon_0 \;\div\; \sqrt{\pi\varepsilon_0}\,\pi\alpha^4\hbar^8 \;=\; \frac{e^2\varepsilon_0}{\sqrt{\pi\varepsilon_0}\,\pi\alpha^4\hbar^8} \;>\; \int\ddagger\int \;>\; \left\{\begin{array}{l} 2.272,843,223,048,749,457,732,376,728,195,2 \\ 1.000,000,006,661,618,217,313,591,999,471,7 \\ 2.272,843,207,907,935,738,835,408,105,450,1 \end{array}\right.$$

$$\sqrt{\pi\varepsilon_0}\,e^4\varepsilon_0 \;\div\; \pi\alpha^4\hbar^6 \;=\; \frac{\sqrt{\pi\varepsilon_0}\,e^4\varepsilon_0}{\pi\alpha^4\hbar^6} \;>\; \int\ddagger\int \;>\; \left\{\begin{array}{l} 9.730,598,001,115,305,996,316,261,554,976,6 \\ 7.940,934,276,075,639,012,859,360,641,827,8 \\ 1.225,371,935,193,009,532,566,580,566,410,1 \end{array}\right.$$

$$\frac{\sqrt{\pi\varepsilon_0}\,e^4\varepsilon_0}{\pi\alpha^4\hbar^6} \;\div\; \frac{\sqrt{\pi\varepsilon_0}\,e^8 c^2\varepsilon_0}{\pi\alpha^6\hbar^8} \;=\; \frac{\alpha^2\hbar^2}{e^4 c^2} \;>\; \iint \;>\; \left\{\begin{array}{l} 7.940,934,276,075,639,012,859,360,641,827,8 \\ 5.922,197,754,666,738,586,193,956,454,205,6 \\ \vee\;\; 1.000,000,000,000,006,537,188,612,504,934,9 \\ \dagger\;\; 1.000,000,000,000,000,000,000,000,000,000,0 \\ 5.922,197,754,666,738,586,193,956,454,205,6 \\ 7.940,934,276,075,587,101,474,238,430,381,5 \end{array}\right.$$

$$\sqrt{\pi\varepsilon_0}\,e^8 c^2\varepsilon_0 \;\div\; \pi\alpha^6\hbar^8 \;=\; \frac{\sqrt{\pi\varepsilon_0}\,e^8 c^2\varepsilon_0}{\pi\alpha^6\hbar^8} \;>\; \int\ddagger\int \;>\; \left\{\begin{array}{l} 5.762,652,563,376,971,982,049,225,428,309,7 \\ 7.940,934,276,075,587,101,474,238,430,381,6 \\ 7.256,894,923,231,724,801,589,587,189,309,4 \end{array}\right.$$

$$\sqrt{8\pi}\,\alpha^{12}\hbar^8 c^9\varepsilon_0^6 \div \sqrt{8\varepsilon_0}\,\pi^6 e^{18}\lambda_p = \frac{\sqrt{8\pi}\,\alpha^{12}\hbar^8 c^9\varepsilon_0^6}{\sqrt{8\varepsilon_0}\,\pi^6 e^{18}\lambda_p} > \int\ddagger\int > \left\{\begin{array}{l} 1.648,130,424,384,889,715,442,993,116,398,3 \\ 1.000,000,008,433,980,347,412,782,131,980,3 \\ 1.648,130,410,484,590,223,442,597,640,913,5 \end{array}\right.$$

$$\frac{\sqrt{8\pi}\,\alpha^{12}\hbar^8 c^9\varepsilon_0^6}{\sqrt{8\varepsilon_0}\,\pi^6 e^{18}\lambda_p} \div \frac{\sqrt{\pi\varepsilon_0}\,\alpha^3 c^{18}\varepsilon_0^5}{\pi^6\hbar\,\lambda_p} = \frac{\sqrt{8\pi}\,\alpha^9\hbar^9\varepsilon_0}{\sqrt{\pi\varepsilon_0}\sqrt{8\varepsilon_0}\,e^{18}c^9} > \iint > \left\{\begin{array}{l} 1.000,000,008,433,980,347,412,782,131,980,3 \\ 9.466,143,283,977,126,171,847,478,873,690,6 \\ \vee\;\; 1.000,000,000,000,006,537,188,612,504,934,9 \\ \dagger\;\; 1.000,000,000,000,000,000,000,000,000,000,0 \\ 9.466,143,283,977,126,171,847,478,873,690,6 \\ 1.000,000,008,433,973,810,224,114,492,568,4 \end{array}\right.$$

$$\sqrt{\pi\varepsilon_0}\,\alpha^3 c^{18}\varepsilon_0^5 \;\div\; \pi^6\hbar\,\lambda_p \;=\; \frac{\sqrt{\pi\varepsilon_0}\,\alpha^3 c^{18}\varepsilon_0^5}{\pi^6\hbar\,\lambda_p} \;>\; \int\ddagger\int \;>\; \left\{\begin{array}{l} 1.349,464,901,828,688,175,636,534,192,738,6 \\ 1.000,000,008,433,973,810,224,114,492,568,4 \\ 1.349,464,890,447,336,631,786,743,216,834,2 \end{array}\right.$$

$$\sqrt{\pi\varepsilon_0}\ \alpha^4 \ \div\ \pi^2 e^{14} c^8 \ =\ \frac{\sqrt{\pi\varepsilon_0}\ \alpha^4}{\pi^2 e^{14} c^8} \ >\ \int \ddagger \int \ > \left\{ \begin{array}{l} 4.729,449,796,671,474,755,163,827,362,779,9 \\ 1.000,000,006,661,624,754,502,248,052,661,1 \\ 4.729,449,765,165,655,124,361,455,340,993,8 \end{array} \right.$$

$$\frac{\sqrt{\pi\varepsilon_0}\ \alpha^4}{\pi^2 e^{14} c^8} \ \div\ \frac{\sqrt{8\pi}\ e^2\varepsilon_0}{\sqrt{8\varepsilon_0}\ \pi^2\alpha^4\hbar^8} \ =\ \frac{\sqrt{\pi\varepsilon_0}\sqrt{8\varepsilon_0}\ \alpha^8\hbar^8}{\sqrt{8\pi}\ e^{16}c^8\varepsilon_0} \ >\ \iint \ > \left\{ \begin{array}{l} 1.000,000,006,661,624,754,502,248,052,661,1 \\ 5.460,096,032,626,260,033,120,061,597,293,5 \\ \text{V}\quad 1.000,000,000,000,006,537,188,612,504,934,9 \\ \dagger\quad 1.000,000,000,000,000,000,000,000,000,000,0 \\ 5.460,096,032,626,260,033,120,061,597,293,5 \\ 1.000,000,006,661,618,217,313,591,999,471,7 \end{array} \right\}$$

$$\sqrt{8\pi}\ e^2\varepsilon_0 \ \div\ \sqrt{8\varepsilon_0}\ \pi^2\alpha^4\hbar^8 \ =\ \frac{\sqrt{8\pi}\ e^2\varepsilon_0}{\sqrt{8\varepsilon_0}\ \pi^2\alpha^4\hbar^8} \ >\ \int \ddagger \int \ > \left\{ \begin{array}{l} 1.139,434,617,339,489,817,350,034,324,075,5 \\ 1.000,000,006,661,618,217,313,591,999,471,7 \\ 1.139,434,609,749,011,463,608,416,182,821,1 \end{array} \right.$$

$$\frac{\alpha^3\varepsilon_0\ \lambda_p}{e^8\hbar} \ \div\ \pi^2\alpha^2\hbar^3 c^3\ N_A \ =\ \frac{\alpha\varepsilon_0\ \lambda_p}{\pi^2 e^8\hbar^4 c^3\ N_A} \ >\ \int \ddagger \int \ > \left\{ \begin{array}{l} 1.000,165,402,709,875,094,497,946,752,503,2 \\ 1.000,000,006,661,624,754,502,248,052,661,4 \\ 1.000,165,396,047,148,533,593,717,207,496,4 \end{array} \right.$$

$$\frac{\alpha\varepsilon_0\ \lambda_p}{\pi^2 e^8\hbar^4 c^3 N_A} \ \div\ \frac{\sqrt{8\pi}\ e^2\varepsilon_0}{\sqrt{8\varepsilon_0}\pi^2\alpha^4\hbar^8} \ =\ \frac{\sqrt{8\varepsilon_0}\ \alpha^5\hbar^4\ \lambda_p}{\sqrt{8\pi}e^{10}c^3\ N_A} \ >\ \iint \ > \left\{ \begin{array}{l} 1.000,000,006,661,624,754,502,248,052,661,1 \\ 9.066,438,696,152,927,731,567,723,977,323,4 \\ \text{V}\quad 1.000,000,000,000,006,537,188,612,504,934,9 \\ \dagger\quad 1.000,000,000,000,000,000,000,000,000,000,0 \\ 9.066,438,696,152,927,731,567,723,977,323,4 \\ 1.000,000,006,661,618,217,313,591,999,471,7 \end{array} \right\}$$

$$\sqrt{8\pi}\ e^2\varepsilon_0 \ \div\ \sqrt{8\varepsilon_0}\ \pi^2\alpha^4\hbar^8 \ =\ \frac{\sqrt{8\pi}\ e^2\varepsilon_0}{\sqrt{8\varepsilon_0}\ \pi^2\alpha^4\hbar^8} \ >\ \int \ddagger \int \ > \left\{ \begin{array}{l} 1.139,434,617,339,489,817,350,034,324,075,5 \\ 1.000,000,006,661,618,217,313,591,999,471,7 \\ 1.139,434,609,749,011,463,608,416,182,821,1 \end{array} \right.$$

$$\varepsilon_0^2 \ \div\ \sqrt{\pi\varepsilon_0}\ \alpha^2\hbar^4 c^2 \ =\ \frac{\varepsilon_0^2}{\sqrt{\pi\varepsilon_0}\ \alpha^2\hbar^4 c^2} \ >\ \int \ddagger \int \ > \left\{ \begin{array}{l} 9.730,598,001,115,305,996,316,261,554,976,6 \\ 7.940,934,276,075,639,012,859,360,641,827,8 \\ 1.225,371,935,193,009,532,566,580,566,410,1 \end{array} \right.$$

$$\frac{\varepsilon_0^2}{\sqrt{\pi\varepsilon_0}\ \alpha^2\hbar^4 c^2} \ \div\ \frac{\sqrt{\pi\varepsilon_0}\ e^8 c^2\varepsilon_0}{\pi\alpha^6\hbar^8} \ =\ \frac{\alpha^4\hbar^4}{e^8 c^4} \ >\ \iint \ > \left\{ \begin{array}{l} 7.940,934,276,075,639,012,859,360,641,827,8 \\ 5.922,197,754,666,738,586,193,956,454,205,6 \\ \text{V}\quad 1.000,000,000,000,006,537,188,612,504,934,9 \\ \dagger\quad 1.000,000,000,000,000,000,000,000,000,000,0 \\ 5.922,197,754,666,738,586,193,956,454,205,6 \\ 7.940,934,276,075,587,101,474,238,430,381,5 \end{array} \right\}$$

$$\sqrt{\pi\varepsilon_0}\ e^8 c^2\varepsilon_0 \ \div\ \pi\alpha^6\hbar^8 \ =\ \frac{\sqrt{\pi\varepsilon_0}\ e^8 c^2\varepsilon_0}{\pi\alpha^6\hbar^8} \ >\ \int \ddagger \int \ > \left\{ \begin{array}{l} 5.762,652,563,376,971,982,049,225,428,309,7 \\ 7.940,934,276,075,587,101,474,238,430,381,6 \\ 7.256,894,923,231,724,801,589,587,189,309,4 \end{array} \right.$$

149

$$\frac{\alpha^3 \varepsilon_0 \lambda_p}{e^8 \hbar} \div \frac{\pi^2 e^2 \alpha \hbar^2 c^4 N_A}{\varepsilon_0} = \frac{\alpha^2 \varepsilon_0^2 \lambda_p}{\pi^2 e^{10} \hbar^3 c^4 N_A} > \int \ddagger \int > \left\{\begin{array}{l} 1.000,165,402,709,875,094,497,946,752,503,2 \\ 1.000,000,006,661,624,754,502,248,052,661,4 \\ 1.000,165,396,047,148,533,593,717,207,496,4 \end{array}\right.$$

$$\frac{\alpha^2 \varepsilon_0^2 \lambda_p}{\pi^2 e^{10} \hbar^3 c^4 N_A} \div \frac{e^2 \varepsilon_0}{\sqrt{\pi\varepsilon_0}\pi\alpha^4\hbar^8} = \frac{\sqrt{\pi\varepsilon_0}\alpha^6\hbar^5\varepsilon_0\lambda_p}{\pi e^{12} c^4 N_A} > \iiint > \left\{\begin{array}{l} 1.000,000,006,661,624,754,502,248,052,661,4 \\ 3.871,300,244,502,524,161,231,643,415,663,6 \\ \varepsilon_0 = 8.854,187,817,620,389,850,536,563,031,710,9 \\ \vee\quad 1.000,000,000,000,006,537,188,612,504,934,3 \\ \dagger\quad 1.000,000,000,000,000,000,000,000,000,000,0 \\ 4.372,281,596,284,182,649,645,495,413,197,5 \\ 1.000,000,006,661,618,217,313,591,999,471,7 \end{array}\right.$$

$$e^2\varepsilon_0 \div \sqrt{\pi\varepsilon_0}\,\pi\alpha^4\hbar^8 = \frac{e^2\varepsilon_0}{\sqrt{\pi\varepsilon_0}\,\pi\alpha^4\hbar^8} > \int \ddagger \int > \left\{\begin{array}{l} 2.272,843,223,048,749,457,732,376,728,195,2 \\ 1.000,000,006,661,618,217,313,591,999,471,7 \\ 2.272,843,207,907,935,738,835,408,105,450,1 \end{array}\right.$$

$$\sqrt{\pi\varepsilon_0}\,e^2\varepsilon_0 \div \pi\alpha^4\hbar^6 = \frac{\sqrt{\pi\varepsilon_0}\,e^2\varepsilon_0}{\pi\alpha^4\hbar^6} > \int \ddagger \int > \left\{\begin{array}{l} 9.730,598,001,115,305,996,316,261,554,976,6 \\ 7.940,934,276,075,639,012,859,360,641,827,8 \\ 1.225,371,935,193,009,532,566,580,566,410,1 \end{array}\right.$$

$$\frac{\sqrt{\pi\varepsilon_0}\,e^2\varepsilon_0}{\pi\alpha^3\hbar^5 c} \div \frac{\sqrt{\pi\varepsilon_0}\,e^8 c^2\varepsilon_0}{\pi\alpha^6\hbar^8} = \frac{\alpha^3\hbar^3}{e^6 c^3} > \iint > \left\{\begin{array}{l} 7.940,934,276,075,639,012,859,360,641,827,8 \\ 4.557,475,660,506,623,129,435,209,322,376,4 \\ \vee\quad 1.000,000,000,000,006,537,188,612,504,934,9 \\ \dagger\quad 1.000,000,000,000,000,000,000,000,000,000,0 \\ 4.557,475,660,506,623,129,435,209,322,376,4 \\ 7.940,934,276,075,587,101,474,238,430,381,5 \end{array}\right.$$

$$\sqrt{\pi\varepsilon_0}\,e^8 c^2\varepsilon_0 \div \pi\alpha^6\hbar^8 = \frac{\sqrt{\pi\varepsilon_0}\,e^8 c^2\varepsilon_0}{\pi\alpha^6\hbar^8} > \int \ddagger \int > \left\{\begin{array}{l} 5.762,652,563,376,971,982,049,225,428,309,7 \\ 7.940,934,276,075,587,101,474,238,430,381,6 \\ 7.256,894,923,231,724,801,589,587,189,309,4 \end{array}\right.$$

$$e^2\varepsilon_0^2 \div \sqrt{\pi\varepsilon_0}\,\alpha^3\hbar^5 c = \frac{e^2\varepsilon_0^2}{\sqrt{\pi\varepsilon_0}\,\alpha^3\hbar^5 c} > \int \ddagger \int > \left\{\begin{array}{l} 2.012,418,077,687,929,991,317,962,034,237,3 \\ 7.940,934,276,075,639,012,859,360,641,827,8 \\ 2.534,233,388,319,207,517,054,120,838,568,3 \end{array}\right.$$

$$\frac{e^2\varepsilon_0^2}{\sqrt{\pi\varepsilon_0}\,\alpha^3\hbar^5 c} \div \frac{\sqrt{\pi\varepsilon_0}\,e^8 c^2\varepsilon_0}{\pi\alpha^6\hbar^8} = \frac{\alpha^3\hbar^3}{e^6 c^3} > \iint > \left\{\begin{array}{l} 7.940,934,276,075,639,012,859,360,641,827,8 \\ 4.557,475,660,506,623,129,435,209,322,376,4 \\ \vee\quad 1.000,000,000,000,006,537,188,612,504,934,9 \\ \dagger\quad 1.000,000,000,000,000,000,000,000,000,000,0 \\ 4.557,475,660,506,623,129,435,209,322,376,4 \\ 7.940,934,276,075,587,101,474,238,430,381,5 \end{array}\right.$$

$$\sqrt{\pi\varepsilon_0}\,e^8 c^2\varepsilon_0 \div \pi\alpha^6\hbar^8 = \frac{\sqrt{\pi\varepsilon_0}\,e^8 c^2\varepsilon_0}{\pi\alpha^6\hbar^8} > \int \ddagger \int > \left\{\begin{array}{l} 5.762,652,563,376,971,982,049,225,428,309,7 \\ 7.940,934,276,075,587,101,474,238,430,381,6 \\ 7.256,894,923,231,724,801,589,587,189,309,4 \end{array}\right.$$

$$\frac{\pi\varepsilon_0}{\sqrt{\pi\varepsilon_0}\ \alpha^3\hbar^3 c} \ \div\ \pi^2\hbar^4 \ =\ \frac{\varepsilon_0}{\sqrt{\pi\varepsilon_0}\ \pi\alpha^3\hbar^7 c} > \int \ddagger \int > \left\{\begin{array}{l} 1.220{,}686{,}778{,}094{,}534{,}904{,}283{,}397{,}451{,}584{,}9 \\ 1.000{,}000{,}006{,}661{,}624{,}754{,}502{,}248{,}052{,}661{,}4 \\ 1.220{,}686{,}769{,}962{,}777{,}700{,}005{,}966{,}313{,}502{,}5 \end{array}\right.$$

$$\frac{\varepsilon_0}{\sqrt{\pi\varepsilon_0}\pi\alpha^3\hbar^7 c} \ \div\ \frac{\sqrt{8\pi}\ e^2\varepsilon_0}{\sqrt{8\varepsilon_0}\pi^2\alpha^4\hbar^8} = \frac{\sqrt{8\varepsilon_0}\ \pi\alpha\hbar}{\sqrt{\pi\varepsilon_0}\sqrt{8\pi}\ e^2 c} > \iint > \left\{\begin{array}{l} 1.000{,}000{,}006{,}661{,}624{,}754{,}502{,}248{,}052{,}661{,}4 \\ 6.434{,}438{,}939{,}006{,}598{,}470{,}018{,}062{,}794{,}553{,}6 \\ \vee\ 1.000{,}000{,}000{,}000{,}006{,}537{,}188{,}612{,}504{,}934{,}9 \\ \dagger\ 1.000{,}000{,}000{,}000{,}000{,}000{,}000{,}000{,}000{,}000{,}0 \\ 6.434{,}438{,}939{,}006{,}598{,}470{,}018{,}062{,}794{,}553{,}6 \\ 1.000{,}000{,}006{,}661{,}618{,}217{,}313{,}591{,}999{,}471{,}7 \end{array}\right.$$

$$\sqrt{8\pi}\ e^2\varepsilon_0 \ \div\ \sqrt{8\varepsilon_0}\ \pi^2\alpha^4\hbar^8 = \frac{\sqrt{8\pi}\ e^2\varepsilon_0}{\sqrt{8\varepsilon_0}\ \pi^2\alpha^4\hbar^8} > \int \ddagger \int > \left\{\begin{array}{l} 1.139{,}434{,}617{,}339{,}489{,}817{,}350{,}034{,}324{,}075{,}5 \\ 1.000{,}000{,}006{,}661{,}618{,}217{,}313{,}591{,}999{,}471{,}7 \\ 1.139{,}434{,}609{,}749{,}011{,}463{,}608{,}416{,}182{,}821{,}1 \end{array}\right.$$

$$\varepsilon_0^2\ \lambda_p m_e \ \div\ \pi\alpha^2\hbar^5\ A_r(e) = \frac{\varepsilon_0^2\ \lambda_p m_e}{\pi\alpha^2\hbar^5 A_r(e)} > \int \ddagger \int > \left\{\begin{array}{l} 9.505{,}449{,}270{,}806{,}645{,}671{,}160{,}092{,}897{,}109{,}5 \\ 7.940{,}934{,}276{,}075{,}639{,}012{,}859{,}360{,}641{,}827{,}6 \\ 1.197{,}019{,}008{,}134{,}163{,}826{,}318{,}379{,}890{,}181{,}4 \end{array}\right.$$

$$\frac{\varepsilon_0^2\ \lambda_p m_e}{\pi\alpha^2\hbar^5 A_r(e)} \ \div\ \frac{\sqrt{\pi\varepsilon_0}e^8 c^2\varepsilon_0}{\pi\alpha^6\hbar^8} = \frac{\alpha^4\hbar^3\varepsilon_0\ \lambda_p m_e}{\sqrt{\pi\varepsilon_0}\ e^8 c^2 A_r(e)} > \iint > \left\{\begin{array}{l} 7.940{,}934{,}276{,}075{,}639{,}012{,}859{,}360{,}641{,}827{,}6 \\ 3.570{,}373{,}133{,}489{,}126{,}091{,}119{,}781{,}312{,}123{,}7 \\ \vee\ 1.000{,}000{,}000{,}000{,}006{,}537{,}188{,}612{,}504{,}934{,}9 \\ \dagger\ 1.000{,}000{,}000{,}000{,}000{,}000{,}000{,}000{,}000{,}000{,}0 \\ 3.570{,}373{,}133{,}489{,}126{,}091{,}119{,}781{,}312{,}123{,}7 \\ 7.940{,}934{,}276{,}075{,}587{,}101{,}474{,}238{,}430{,}381{,}5 \end{array}\right.$$

$$\sqrt{\pi\varepsilon_0}\ e^8 c^2\varepsilon_0 \ \div\ \pi\alpha^6\hbar^8 \ =\ \frac{\sqrt{\pi\varepsilon_0}\ e^8 c^2\varepsilon_0}{\pi\alpha^6\hbar^8} > \int \ddagger \int > \left\{\begin{array}{l} 5.762{,}652{,}563{,}376{,}971{,}982{,}049{,}225{,}428{,}309{,}7 \\ 7.940{,}934{,}276{,}075{,}587{,}101{,}474{,}238{,}430{,}381{,}6 \\ 7.256{,}894{,}923{,}231{,}724{,}801{,}589{,}587{,}189{,}309{,}4 \end{array}\right.$$

$$\frac{\alpha^4\hbar^4\varepsilon_0}{\pi^3} \ \div\ \frac{\sqrt{\pi\varepsilon_0}\ e^{14}\hbar^4 c^8}{\pi^2} \ =\ \frac{\alpha^4\varepsilon_0}{\sqrt{\pi\varepsilon_0}\pi e^{14}c^8} > \int \ddagger \int > \left\{\begin{array}{l} 1.001{,}532{,}213{,}616{,}395{,}767{,}283{,}384{,}936{,}299{,}6 \\ 1.000{,}000{,}006{,}661{,}624{,}754{,}502{,}248{,}052{,}661{,}4 \\ 1.001{,}532{,}206{,}944{,}564{,}025{,}070{,}208{,}920{,}538{,}8 \end{array}\right.$$

$$\frac{\alpha^4\varepsilon_0}{\sqrt{\pi\varepsilon_0}\ \pi e^{14}c^8} \ \div\ \frac{e^2\varepsilon_0}{\sqrt{\pi\varepsilon_0}\ \pi\alpha^4\hbar^8} \ =\ \frac{\alpha^8\hbar^8}{e^{16}c^8} > \iint > \left\{\begin{array}{l} 1.000{,}000{,}006{,}661{,}624{,}754{,}502{,}248{,}052{,}661{,}1 \\ 1.230{,}075{,}082{,}737{,}603{,}011{,}435{,}284{,}283{,}985{,}1 \\ \vee\ 1.000{,}000{,}000{,}000{,}006{,}537{,}188{,}612{,}504{,}934{,}9 \\ \dagger\ 1.000{,}000{,}000{,}000{,}000{,}000{,}000{,}000{,}000{,}000{,}0 \\ 1.230{,}075{,}082{,}737{,}603{,}011{,}435{,}284{,}283{,}985{,}1 \\ 1.000{,}000{,}006{,}661{,}618{,}217{,}313{,}591{,}999{,}471{,}7 \end{array}\right.$$

$$e^2\varepsilon_0 \ \div\ \sqrt{\pi\varepsilon_0}\ \pi\alpha^4\hbar^8 \ =\ \frac{e^2\varepsilon_0}{\sqrt{\pi\varepsilon_0}\ \pi\alpha^4\hbar^8} > \int \ddagger \int > \left\{\begin{array}{l} 2.272{,}843{,}223{,}048{,}749{,}457{,}732{,}376{,}728{,}195{,}2 \\ 1.000{,}000{,}006{,}661{,}618{,}217{,}313{,}591{,}999{,}471{,}7 \\ 2.272{,}843{,}207{,}907{,}935{,}738{,}835{,}408{,}105{,}450{,}1 \end{array}\right.$$

$$\pi\alpha^6\hbar^8 \div \sqrt{\pi\varepsilon_0}\, e^8 c^2 \varepsilon_0 = \frac{\pi\alpha^6\hbar^8}{\sqrt{\pi\varepsilon_0}\, e^8 c^2 \varepsilon_0} > \int \ddagger \int > \left\{ \begin{array}{l} 7.256,894,923,231,724,801,589,587,189,309,4 \\ 1.259,297,666,035,085,735,991,190,862,792,4 \\ 5.762,652,563,376,971,982,049,225,428,309,7 \end{array} \right.$$

$$\frac{\pi\alpha^6\hbar^8}{\sqrt{\pi\varepsilon_0}\, e^8 c^2 \varepsilon_0} \div \frac{\pi\alpha^4\hbar^6}{\sqrt{\pi\varepsilon_0}\, e^4 \varepsilon_0} = \frac{\alpha^2\hbar^2}{e^4 c^2} > \iint > \left\{ \begin{array}{l} 1.259,297,666,035,085,735,991,190,862,792,4 \\ 5.922,197,754,666,738,586,193,956,454,205,6 \\ \vee\ 1.000,000,000,000,006,537,188,612,504,934,9 \\ \dagger\ 1.000,000,000,000,000,000,000,000,000,000,0 \\ 5.922,197,754,666,738,586,193,956,454,205,6 \\ 1.259,297,666,035,077,503,724,828,704,241,3 \end{array} \right.$$

$$\pi\alpha^4\hbar^6 \div \sqrt{\pi\varepsilon_0}\, e^4 \varepsilon_0 = \frac{\pi\alpha^4\hbar^6}{\sqrt{\pi\varepsilon_0}\, e^4 \varepsilon_0} > \int \ddagger \int > \left\{ \begin{array}{l} 1.225,371,935,193,009,532,566,580,566,410,1 \\ 1.259,297,666,035,077,503,724,828,704,241,3 \\ 9.730,598,001,115,305,996,316,261,554,976,6 \end{array} \right.$$

$$\pi\alpha^6\hbar^8 \div \sqrt{\pi\varepsilon_0}\, e^8 c^2 \varepsilon_0 = \frac{\pi\alpha^6\hbar^8}{\sqrt{\pi\varepsilon_0}\, e^8 c^2 \varepsilon_0} > \int \ddagger \int > \left\{ \begin{array}{l} 7.256,894,923,231,724,801,589,587,189,309,4 \\ 1.259,297,666,035,085,735,991,190,862,792,4 \\ 1.259,297,666,035,077,503,724,828,704,241,4 \end{array} \right.$$

$$\frac{\pi\alpha^6\hbar^8}{\sqrt{\pi\varepsilon_0}\, e^8 c^2 \varepsilon_0} \div \frac{\sqrt{\pi\varepsilon_0}\,\alpha^6\hbar^8}{e^8 c^2 \varepsilon_0^2} = \int \dagger \int > \int > \left\{ \begin{array}{l} 1.259,297,666,035,085,735,991,190,862,792,4 \\ \vee\ 1.000,000,000,000,006,537,188,612,504,934,8 \\ \dagger\ 1.000,000,000,000,000,000,000,000,000,000,0 \\ 1.259,297,666,035,077,503,724,828,704,241,4 \end{array} \right.$$

$$\sqrt{\pi\varepsilon_0}\,\alpha^6\hbar^8 \div e^8 c^2 \varepsilon_0^2 = \frac{\sqrt{\pi\varepsilon_0}\,\alpha^6\hbar^8}{e^8 c^2 \varepsilon_0^2} > \int \ddagger \int > \left\{ \begin{array}{l} 3.852,567,560,358,045,371,299,054,355,542,9 \\ 1.259,297,666,035,077,503,724,828,704,241,4 \\ 3.059,298,579,094,430,572,855,128,522,780,7 \end{array} \right.$$

$$\pi c^4 \varepsilon_0^2 K_B^2 \div \sqrt{\pi\varepsilon_0}\, e^8 \alpha^4 \hbar^4 = \frac{\pi c^4 \varepsilon_0^2 K_B^2}{\sqrt{\pi\varepsilon_0}\, e^8 \alpha^4 \hbar^4} > \int \ddagger \int > \left\{ \begin{array}{l} 3.792,246,709,501,433,224,270,334,799,396,1 \\ 6.097,799,255,893,838,312,759,867,034,636,4 \\ 6.219,041,576,083,421,692,332,093,358,918,5 \end{array} \right.$$

$$\frac{\pi c^4 \varepsilon_0^2 K_B^2}{\sqrt{\pi\varepsilon_0}\, e^8 \alpha^4 \hbar^4} \div \frac{\sqrt{\pi\varepsilon_0}\, c^4 \varepsilon_0 K_B^2}{e^8 \alpha^4 \hbar^4} > \int \dagger \int > \iint > \left\{ \begin{array}{l} 6.097,799,255,893,838,312,759,867,034,636,5 \\ \vee\ \sqrt{\pi\varepsilon_0}\sqrt{2c} = 4.083,895,782,204,487,460,490,369,1 \\ \vee\ \sqrt{8c}\,\pi c\varepsilon_0 = 4.083,895,782,204,487,460,490,369,1 \\ \dagger\ 1.000,000,000,000,000,000,000,000,000,000,0 \\ 1.493,132,925,297,678,752,736,396,006,964,7 \end{array} \right.$$

$$\sqrt{\pi\varepsilon_0}\, c^4 \varepsilon_0 K_B^2 \div e^8 \alpha^4 \hbar^4 = \frac{\sqrt{\pi\varepsilon_0}\, c^4 \varepsilon_0 K_B^2}{e^8 \alpha^4 \hbar^4} > \int \ddagger \int > \left\{ \begin{array}{l} 2.273,773,924,767,693,241,653,737,851,987,9 \\ 6.097,799,255,893,838,312,759,867,034,636,4 \\ 3.728,843,520,996,486,004,389,516,040,711,3 \end{array} \right.$$

These navigation tables are just a few examples. The complete listing is still in the process of being published.

Ingram Content Group UK Ltd.
Milton Keynes UK
UKHW050619170523
421848UK00003B/45